Contributions to Management Science

More information about this series at http://www.springer.com/series/1505

Nadi Serhan Aydın

Financial Modelling with Forward-looking Information

An Intuitive Approach to Asset Pricing

Springer

Nadi Serhan Aydın
Ankara, Turkey

ISSN 1431-1941 ISSN 2197-716X (electronic)
Contributions to Management Science
ISBN 978-3-319-57146-1 ISBN 978-3-319-57147-8 (eBook)
DOI 10.1007/978-3-319-57147-8

Library of Congress Control Number: 2017942803

Printed on acid-free paper

This Springer imprint is published by Springer Nature
The registered company is Springer International Publishing AG
The registered company address is: Gewerbestrasse 11, 6330 Cham, Switzerland

To my parents

Foreword

The purpose of this book, *Financial Modelling with Forward-Looking Information: An Intuitive Approach to Asset Pricing*, is to deeply inquire, holistically reflect on, and practically expose the current and emerging concept of information-based modelling to the areas of financial market microdynamics and asset pricing with real-time signals. During the previous decades, the analytical tools and the methodological toolbox of applied and financial mathematics, and of statistics, have gained the attention of numerous researchers and practitioners from all over the world, providing a strong impact also in economics and finance. Here, the notions of futuristic information on asset fundamentals and informational disparities among market participants are turning out to be key issues from an integrated perspective, and they are closely connected with further areas such as financial signal processing, market microstructure, agent-based modelling, and early detection of financial bubbles and liquidity squeezes.

This book seeks to reassess and revitalize, amid ongoing structural problems in financial markets, the role of information through a fundamental approach that can be used for pricing a broad spectrum of financial and insurance contracts. The approach focuses on an intuitive, yet theoretically robust, framework for integrating financial information flows, which is also known as the *Brody, Hughston and Macrina* framework. This book could become a helpful compendium for decision-makers, researchers, as well as graduate students and practitioners in quantitative finance who aim to go beyond conventional approaches to financial modelling.

The author of this book is both an academic and practitioner in the area of applied financial mathematics, with considerable international research experience. He uses the state-of-the-art model-based strong methods of mathematics as well as the less model-based, more data-driven algorithms—often called as heuristics and model-free—which are less rigorous mathematically and released from firm calculus in order to integrate data-led approaches with a view to efficiently coping with hard problems. Today, labeled by names like Statistical or Deep Learning and Adaptive Algorithms, and by Operational Research and Analytics, model-free and model-based streamlines of traditions and approaches meet and exchange in various centers of research, at important congresses, and in leading projects and agendas in all over the world. The herewith joint intellectual enterprise aims to benefit from synergy effects, to commonly advance scientific progress and to provide a united and committed service to the solution of urgent real-life challenges.

To the author of this valuable book, **Dr. Nadi Serhan Aydın**, I extend my heartily appreciation and gratitude for having shared his devotion, knowledge, and vision with the academic community and mankind. I am very thankful to the publishing house ***Springer***, and the editorial team around ***Dr. Christian Rauscher*** thereat, for having ensured and made become reality a premium work of a high-standard academic and applied importance, and a future promise of a remarkable impact for the world of tomorrow.

Now, I wish all of you a lot of joy in reading this interesting work, and I hope that a great benefit is gained from it both personally and societally.

Middle East Technical University
Ankara, Turkey
March 2017

Gerhard-Wilhelm Weber

Acknowledgments

At the outset, I would like to cordially thank Prof. Gerhard-Wilhelm Weber, my supervisor at Middle East Technical University (METU), who has not only been a great scientific mentor but also an excellent collaborator and friend. I would also like to thank Prof. Anthony G. Constantinides, my co-supervisor at Imperial College London (ICL), for both engaging me in the extremely vibrant research environment of ICL and sharing with me his intriguing ideas which have no doubt enriched this work.

Besides, I am also grateful to the members of the Thesis Monitoring Committee at METU, namely, Assoc. Prof. Azize Hayfavi and Assoc. Prof. Yeliz Yolcu Okur, for their insightful comments during our regular follow-up meetings at the Institute of Applied Mathematics (IAM).

Among many others, I owe special thanks to Edward Hoyle, PhD (ICL), with whom I had a series of inspiring conversations in London that have influenced this work. I also thank Arta Babaee, PhD (ICL), and Pedro Rodrigues, PhD (ICL), who helped me delve into the emerging area of Financial Signal Processing (FSP).

The ICL staff were extremely helpful in providing me with access to ICL's exclusive data sources and other research facilities. So, I am grateful to them and, in particular, Jason Murray from the Business School.

This research was supported by the Turkish Scientific and Technological Research Council (TÜBİTAK) under its doctoral scholarship (no. 2211) and international doctoral research scholarship (no. 2214) programs.

Finally, I would like to dedicate this work to my dear parents for they have shared with me all the joys and sorrows of this life, including of this work.

Contents

List of Figures

List of Tables

List of Abbreviations

a.e.	Almost every
CHDE	Confluent hypergeometric differential equation
DSGE	Dynamic stochastic general equilibrium
DSP	Digital signal processing
FSP	Financial signal processing
GBM	Geometric Brownian motion
OU	Ornstein–Uhlenbeck
$\mathbb{Z}^+, \mathbb{Z}^-$	Set of positive, negative integers
P&L	Profit-and-loss
$\mathbb{R}$	Set of real numbers
$\mathbb{C}$	Set of complex numbers
s.t.	Subject to
w.r.t.	With respect to
w.l.o.g.	Without loss of generality

Chapter 1
Introduction

The raison d'être of the markets we study is to support information-based trading. Yet, there is a fundamental conflict between how efficiently markets spread information and the incentives to acquire it. This is something conventional stochastic models and, particularly, the way their information content is structured tend to oversimplify. As such, the notion that "there is a universal market filtration" also seems to be unrealistic. What counts, for market efficiency, is that, in practice, investors have access to different levels of information and with varying ease. This calls for a broader view of market efficiency which takes into account the amount and pace of such access. Nevertheless, by exchanging information through highly frequent trades, market participants are able to maintain a law of reasonable price range, if not a law of one price.

Complications related to construction of an information flow are generally bypassed through the concept of "natural filtration" $\mathcal{F}$, whereas the essential point is that all relevant information is contained in, and therefore, can be extracted from, the past trajectory. Yet, little is known about the structure of this filtration. It is not clear, for example, why a stochastic driver should be regarded as to contain all relevant information about the "fundamental" value rather than noise. The filtration generated by this random process is also pre-imposed on the future evolution of the "fundamental" value. We emphasise here the word "fundamental" to reflect the notion that an asset's future is not necessarily determined by its past, but also its future prospects.

In this book, we focus on a concept where some of the aforementioned problems are sought to be addressed. Market participants get noisy signals ξ on the future convenience dividends of an asset directly, or market factors which affect them.[1] When combined together, the signals ξ form the "all-wise" filtration. The informational diversity thus naturally stems from the fact that either ξs might differ in quality (i.e., in their signal-to-noise) or agents might vary in their capacity to

[1]We refer as convenience dividends to any material benefit drawn from holding the asset.

N.S. Aydın, *Financial Modelling with Forward-looking Information*, Contributions to Management Science, DOI 10.1007/978-3-319-57147-8_1

interpret the same signal (cf. [1]). In this structure, a subset of all available signals could determine the filtration of the agent rather more explicitly. As a result, the question of how real-time information flow dynamics can be satisfactorily imitated, as well as its implications for asset pricing and market microstructure, need to be brought more under spotlight.

Assume we know a priori that a business—with two possible outcomes—will default at maturity. Had it not failed, the business would pay one unit to its investors. The only thing the agents know when the business started is that the two outcomes would have even chances. Therefore, it is natural for them all to value the business at an initial price of 0.5. But, once they start to get rumours about the health of the business through different sources, the situation will change. As some of these sources will be more reliable than others, revealing the true status of the business at a faster pace, investors will start to differ in their judgements about the real value of the latter and, if allowed, try to exploit that information. One who has access to a fast-track signal source will uncover quickly what the real outcome will be, constantly trimming the value to the asset, whereas the ones with access to less superior information sources will have to wait longer periods to see what is happening, putting any bet they make during that time at the risk of being exploited by others (cf. Fig. 1.1, left panel). Although it does not mean that the faster signals will get the investor a more realistic value judgement at all instances and throughout the horizon, on average, they will do so (Fig. 1.1, right panel).

One benefit, inter alia, of working with signals rather than their aggregate (e.g., price) directly is that the signals on certain factors can be both more accessible and predictable. Consider the likelihood of a regular policy decision on interest rates, an intervention on the value of a currency, positive earnings announcement, a merger or acquisition, certain regulatory changes, the resigning of a company's top management, or a candidate winning the approaching national poll. It may be

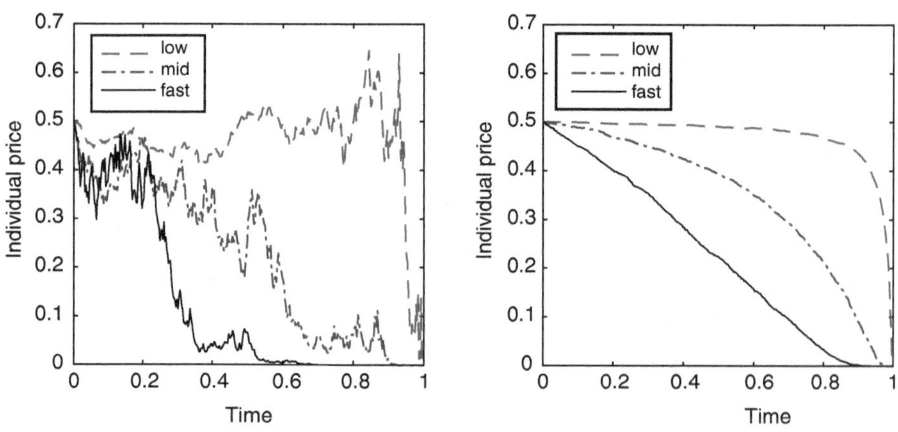

Fig. 1.1 Differences in value judgements based on individual information sources. The real outcome is set to default with a priori probability of 0.5

much easier to collect signals on the outcome of ongoing discussions on a regulatory change that would impact the way a company is running, judge the reliability of signal sources, and determine the relative importance of that policy change against other possible factors, than to focus on that company's equity performance.

Needless to say that the ideas presented above are not all new. Yet, the literature on the dynamics of financial information flow is considerably scarce, as compared to that on heterogeneous information (which is also empowered by the recent advances in methods such as the Malliavin calculus (cf. [8])) and stochastic filtering (with ongoing emphasis on generalisations to nonlinear systems, and particle methods (cf. [2, 7])), which can be seen natural extensions of the present framework. In this book, we aim to introduce the framework which was originally developed in [3] and extend it in different directions.

- Accordingly, the next chapter, i.e., Chap. 2, assembles some fundamental properties of random bridge processes and justifies their use in modelling forward-looking financial information. Although this chapter is essentially based on [3, 5] and [4], it contributes to the existing literature by recovering the necessary properties of the signal-based framework in a much greater detail, and presenting a useful information-theoretic analysis to quantify the information component.
- Chapter 3 introduces an interactive market setup where agents receive variegated information. This chapter, which is inspired by the remarks of authors in [6], is a significant addition to the literature on equilibrium with long-lived information. It not only vividly illustrates some interesting price discovery dynamics in the presence of heterogeneous information through numerical analysis, but also explores optimal strategies to exploit differential information by analytically characterising ex-ante gains from trade.
- Chapter 4 puts the signal-based framework to practical use by introducing a slightly modified version of the signal process and making a particular choice for real-time signals. To the best of the author's knowledge, this is the first such attempt, with results having significant implications for harnessing the signal-based framework in a real-world setting. We also contribute the literature by presenting a crisp formula for the signal-based price.

Finally, Chap. 5 concludes with a brief outlook, and some remarks on the contemporary area of Financial Signal Processing (FSP).

Throughout the book, we may interchange between the terms "dividends" and "cashflows", as well as "agents" and "investors"—which is of no harm. However, a distinction has to be made at the outset between an "investor" and a "trader." In the present context, all market participants are "investors" who make their decisions on the basis of long-term targets, more due diligence and a proper analysis of fundamental factors; while "traders" will not necessarily do so.

References

1. Admati A, Pfleiderer P (1988) A theory of intraday patterns: volume and price variability. Rev Financ Stud 1(1):3–40
2. Bain A, Crisan D (2009) Fundamentals of stochastic filtering. Stochastic modelling and applied probability, vol 60, 1st edn. Springer, New York
3. Brody D, Hughston LP, Macrina A (2007) Beyond hazard rates: a new framework for credit-risk modelling. Advances in mathematical finance. Applied and numerical harmonic analysis, chapter III. Birkhäuser, Boston, pp 231–257
4. Brody D, Hughston LP, Macrina A (2008) Dam rain and cumulative gain. Proc Math Phys Eng Sci 464(2095):1801–1822
5. Brody D, Hughston L, Macrina A (2008) Information-based asset pricing. Int J Theoret Appl Financ 11(1):107–142
6. Brody D, Hughston L, Macrina A (2011) Modelling information flows in financial markets. Advanced mathematical methods for finance. Springer, Berlin, pp 133–153
7. Kallianpur G (1980) Stochastic filtering theory. Springer, New York
8. Kieu A, Oksendal B, Yolcu-Okur Y (2013) A Malliavin calculus approach to general stochastic differential games with partial information. In: Malliavin calculus and stochastic analysis. Springer proceedings in mathematics & statistics, vol 34, chapter 5. Springer, New York, pp 489–510

Chapter 2
The Signal-Based Framework

The flow of forward-looking information through signals is essential for the smooth operation of the highly complex financial market engine and it is the most fundamental input to the pricing of any type of asset. The market agents, both human and non-human, on the other hand, are signal processors who continuously mine for and interpret these signals to extract information.

In what follows, we lay out the basic characteristics of the information-based framework which was first introduced in [8] as a new way of modelling credit risk and, later on, applied to a broad spectrum of issues in financial mathematics, including the valuation of insurance contracts based on the cumulative gain process in [9], modeling of defaultable bonds in [30] (as an extension of [8] to stochastic interest rates), general asset pricing in [10], pricing of inflation-linked assets in [20], and modelling of asymmetric information and insider trading in [11], before it was generalised to a wider class of Lévy information processes in [18] for valuing credit-risky bonds, vanilla and exotic options, and non-life insurance liabilities. This method was used, in [6], to aggregate individual risk aversion dynamics to form a market pricing kernel, in [25], to price credit-risky assets that may include random recovery upon default, in [26], to introduce an extension of the theory towards an analysis of information blockages and activations, as well as information-switching dynamics, in [13], to introduce a general framework for signal processing with Lévy information, in [33], to value storable commodities and associated derivatives and, most recently, in [7], to obtain a stochastic volatility model based on random information flow, and in [2], to produce estimates of bankruptcy time.

However, we distinguish the present analysis from another particular strand of literature which looks at information dissemination and epidemics in networks with certain topological properties, with applications to finance (see, e.g., [3, 5, 14, 15, 23]).

N.S. Aydın, *Financial Modelling with Forward-looking Information*, Contributions to Management Science, DOI 10.1007/978-3-319-57147-8_2

2.1 Modelling Information Flow

The information-based approach stems naturally from the dynamic nature of information. Information is revealed at some pace and it is not pure all the time. There is normally little or no rumour about an asset's future value when there is a significant time frame until its maturity; the beliefs are most diverse around the midway through the lifetime of the asset when the rumours intensify; there is a growing consensus, as the asset approaches its maturity, on how things will turn out; and, finally, the true value becomes known.[1] Bridge processes indeed have some nice properties to imitate this behaviour. Consider a Brownian bridge process defined over the period $[0, T]$[2] which takes on values 0 and z at times 0 and T, respectively:

$$\beta_{0T}^{[0,z]}(t) := B_t - \frac{t}{T}(B_T - z). \tag{2.1}$$

with B_t being a Brownian motion. The bridge process in Eq. (2.1) is a standard Brownian bridge with a deterministic drift. Let z represent the true value at time T of a random quantity Z_T that adheres to the a priori marginal density $f_{Z_T}(z)$, i.e., $Z_T(\omega) = z$. Rearranging the terms of Eq. (2.1) yields a random bridge process:

$$\beta_{0T}^{[0,Z_T]}(t) = \frac{t}{T}Z_T + B_t - \frac{t}{T}B_T = \frac{t}{T}Z_T + \beta_{0T}^{[0,0]}(t), \tag{2.2}$$

where $\beta_{0T}^{[0,0]}(t)$ is a standard Brownian bridge, representing the 'pure noise', that adheres to the law $\mathcal{N}(0, (t\kappa_t^{-1})^{1/2})$ with $\kappa_t := T/(T-t)$.[3] The first part $(t/T)Z_T$, on the other hand, is the 'hidden truth' about the future value of the random variable Z_T (in the sense that it is concealed by noise). The term $1/T$, in this case, governs the overall speed of revelation of true information about the actual value of Z_T.

Definition 2.1 The process $\beta_{0T}^{[0,Z_T]}(t)$ is a 'Brownian random bridge' if:

- Its terminal value $\beta_{0T}^{[0,Z_T]}(T)$ has the marginal law υ which admits density $p(z)$, i.e., $\upsilon(\mathrm{d}z) = p(z)\mathrm{d}z$.
- There exists a Gaussian process $(G_t)_{0\le t\le T}$ with density $g_t(y)$ for all $t \in [0, T]$, and υ concentrates mass where $0 < g_T(z) < \infty$ for υ-almost-every z.

---

[1] Zero-noise at initial date is still intuitive since single point will have no prediction power.

[2] See [2] for bridges on a random intervals $[0, \tau]$.

[3] The part $\mathbb{E}[\beta_t] = 0$ is indeed trivial, whereas $\mathbb{V}[\beta_t] = \mathbb{E}\left[B_t^2 - \frac{t}{T}B_tB_T + \frac{t^2}{T^2}B_T^2\right] = t - 2\frac{t^2}{T} + \frac{t^2}{T} = t\kappa_t^{-1}$.

- Furthermore,

$$\mathbb{Q}\left[\beta_{0T}^{[0,Z_T]}(t_1) \leq y_1, \ldots, \beta_{0T}^{[0,Z_T]}(t_l) \leq y_l \middle| \beta_{0T}^{[0,Z_T]}(T) = z\right]$$
$$= \mathbb{Q}\left[G_{t_1} \leq y_1, \ldots, G_{t_l} \leq y_l \middle| G_T = z\right] \tag{2.3}$$

for every $l \in \mathbb{Z}^+$, increasing $(t_1, \ldots, t_l) \in [0, T]$, $(y_1, \ldots, y_l) \in \mathbb{R}^l$, and υ-almost-every z.[4]

Thus, a Brownian random bridge is identical in law to a Brownian motion conditioned to have the a priori law of Z_T at time T. Indeed, one can define $X_T := Z_T/(\sigma T)$ in Eq. (2.2) by introducing a more general parameter, say σ (or, alternatively, σ_t) instead of $1/T$. This enables us to introduce the signal process ξ_t (or, the information process in the sense of [8]):

$$\xi_t = \sigma t X_T + \beta_t. \tag{2.4}$$

where (and, henceforth) $\beta_t := \beta_{0T}^{[0,0]}(t)$. In other words, σ will be gauging the ratio of true signal to noise (henceforth, just 'signal-to-noise'). This particular way of defining the information flow, in fact, distinguishes the current framework from a large class of asymmetric information models, where, as in [22], a bulk of information is assumed to arrive instantly at the beginning of the trading period, or, as in [1], the arrival pattern of information is found to be irrelevant to trading strategies of agents. We also set β_t and X_T to be independent: $\beta_t \perp\!\!\!\perp X_T$. We note that, hereafter, the signal ξ_t will be regulating the information flow.

We also remark that Eq. (2.4) is not the only way to represent information flow. Some other forms have also been considered in the literature with slightly different characteristics, such as $\xi_t = tX_T + \beta_t$ (cf. [6]), and $\xi_t = (t/T)X_T + \sigma\beta_t$ or $\xi_t = (t/T)X_T + \beta_t$ (cf. [19]).

More formally, we define a probability space $(\Omega, \mathcal{F}, \mathbb{Q})$, on which the filtration $(\mathcal{F}_t^{\xi})_{t\in[0,T]}$ will be constructed. Here, $\mathbb{Q}$, i.e., the risk-neutral measure, is assumed to exist. The default measure is set to $\mathbb{Q}$ throughout the book, if not stated otherwise. For simplicity, we assume that the asset under consideration is of predetermined maturity, i.e., the cashflow will be generated, and the related information process will expire, at a pre-known time T. The filtration $\mathcal{F}_t^{\xi}$, which is assumed to be generated directly by $(\xi_s)_{0\leq s\leq t}$, is given by:

$$\mathcal{F}_t^{\xi} = \{\sigma(\xi_s) : 0 \leq s \leq t < T\}. \tag{2.5}$$

[4]See [18] for a definition of Lévy random bridge instead.

We are now in a position to work out, with respect to the available information $\mathcal{F}_t^{\xi}$, the value S_t and dynamics $\mathrm{d}S_t$ of an asset which generates a cashflow $\phi_T = \phi(X_T)$ at time T for some invertible function ϕ. The value S_t, $0 \leq t < T$, is given by

$$S_t = \mathbf{1}_{\{t<T\}} e^{-r(T-t)} \mathbb{E}\left[\phi_T | \mathcal{F}_t^{\xi}\right],$$

$$\text{(or, simply)} = \mathbf{1}_{\{t<T\}} e^{-r(T-t)} \phi_t, \tag{2.6}$$

where ϕ_t, $\phi_t(X_T)$, $\mathbb{E}[\phi_T|\mathcal{F}_t^{\xi}]$ are all equivalent, and r is the money market rate. Also not to mention that the asset goes ex-dividend at T, i.e., immediately after the dividend is paid, should the asset's maturity be longer than T and should there be other dividends to be paid.

The quantities X_T and $\phi(X_T)$ are measurable with respect to $\mathcal{F}_T^{\xi}$, but not necessarily w.r.t. $\mathcal{F}_t^{\xi}$, $t < T$. On an important note, we remark that β_t, i.e., the pure noise, is not measurable w.r.t. $\mathcal{F}_t^{\xi}$, meaning that it is not directly accessible to market agents. Thus, an agent, although he observes ξ_t, cannot separate true signal from noise until time T.

Note that the expectation in Eq. (2.6) is conditioned, as we understand from Eq. (2.5), on the entire path of ξ_t, which renders it difficult to handle. Therefore, verifying that the information process ξ_t satisfies the Markov property could bring a great deal of simplification to the construction of price dynamics. In [18], it is indeed shown that Lévy bridges, and Lévy random bridges alike, satisfy the Markov property. Here we verify the latter for Brownian random bridges.

Proposition 2.1 *The information process* $(\xi_t)_{0\leq t\leq T}$, *as defined in Eq.* (2.4), *is conditionally Markovian.*

Proof (See an Alternative Proof in [24]) We set $\kappa_t = T/(T-t)$ here and, whenever appropriate, throughout the text. Let ξ_t be intrinsically pinned to an unknown value $X_T = x$. Defining B_t as a Brownian motion, we can indeed express the signal process ξ_t as

$$\sigma t x + \kappa_t^{-\frac{1}{2}} B_t \quad \text{or} \quad \sigma t x + \kappa_t^{-\frac{1}{2}} \int_0^t \mathrm{d}B_s. \tag{2.7}$$

One can verify that these are identical to

$$\xi_t = \sigma t x + (T-t) \int_0^t \frac{\mathrm{d}B_s}{T-s}, \tag{2.8}$$

which, in turn, implies

$$
\begin{aligned}
\mathrm{d}\xi_t &= \left(\sigma x - \int_0^t \frac{\mathrm{d}B_s}{T-s}\right)\mathrm{d}t + (T-t)\frac{\mathrm{d}B_t}{T-t} \\
&= \left(\sigma x - \frac{\xi_t - \sigma t x}{(T-t)}\right)\mathrm{d}t + \mathrm{d}B_t \\
&= (\sigma x - \xi_t/T)\,\kappa_t \mathrm{d}t + \mathrm{d}B_t.
\end{aligned} \tag{2.9}
$$

Equations (2.8) and (2.9) indeed follow from two other well-known representations of bridges (see, e.g., [28]). Equation (2.9), on the other hand, directly implies that, given $X_T = x$, ξ_t is a Markov process with respect to its own filtration, i.e.,

$$
\mathbb{E}[h(\xi_t)|\sigma\,(\xi_r)_{r\le s}] = \mathbb{E}[h(\xi_t)|\sigma(\xi_s)] \quad (s \le t), \tag{2.10}
$$

for any x, and any measurable, finite-valued function h (cf. [28]). □

Proposition 2.1 leads to a significant reduction in the complexity of calculating the expectation in Eq. (2.6). The latter expectation can now be written, again, for the single dividend as

$$
S_t = \mathbf{1}_{\{t<T\}} e^{-r(T-t)} \mathbb{E}\left[\phi_T|\xi_t\right], \tag{2.11}
$$

or, when the payoff has a continuous density, as

$$
S_t = \mathbf{1}_{\{t<T\}} e^{-r(T-t)} \int_{\mathbb{X}} \phi(x)\pi_t(x)\mathrm{d}x. \tag{2.12}
$$

Here, the posterior density $\pi_t(x) := p(x|\xi_t)$ is given by

$$
\pi_t(x) = \frac{\mathrm{d}}{\mathrm{d}x}\mathbb{Q}(X_T \le x|\xi_t). \tag{2.13}
$$

To restore S_t and its dynamics, apparently, we need to work out π_t, the posterior density. Using Bayesian inference, π_t can be written as

$$
\begin{aligned}
\pi_t(x) &= \frac{p(x)p(\xi_t|x)}{\displaystyle\int_{\mathbb{X}} p(y)p(\xi_t|y)\mathrm{d}y} \\
&= \frac{p(x)p(\xi_t|x)}{\displaystyle\int_{\mathbb{X}} p(\xi_t)\mathrm{d}y} \quad (x \in \mathbb{X}),
\end{aligned} \tag{2.14}
$$

where $\mathbb{X}$ is the support of X_T, $p(x)$ the a priori probability density of X_T, and $p(\xi_t|x)$ the likelihood (i.e., compatibility of the signal ξ_t given the measurement x). We note that the procedure in Eq. (2.14) is similar to a Kalman [21] filtering operation in which a transition step based on $p(x|\xi_s)$ and $p(\xi_t|\xi_s)$ also takes place before the measurement update $p(x|\xi_t)$ (see, e.g., [4]).

Here, we find it useful to state a dynamical consistency property satisfied by ξ_t.

Proposition 2.2 *The process ξ_t is dynamically consistent, meaning that, if we store the information transmitted by ξ_s, $s \in [0, T]$, in $\pi_s(x)$ and, then, re-initialise it at time s as ξ'_t, $s \le t \le T$, updating also its flow rate to σ', then $\pi_t(x)$ can be written in terms of $\pi_s(x)$ (i.e., the new prior) as follows*

$$\pi_t(x) = \frac{\pi_s(x)p(\xi'_t|x)}{\int_{\mathbb{X}} \pi_s(y)p(\xi'_t|y)\mathrm{d}y} \quad (s \le t \le T), \tag{2.15}$$

where

$$\xi'_t = \xi_t - \frac{T-t}{T-s}\xi_s = \sigma'(t-s)X_T + \beta'_t, \tag{2.16}$$

with $\sigma' = \sigma T/(T-s)$, and β'_t being a standard Brownian bridge over $[s, T]$ (see a time-varying information flow version in [10]).

Proof Calculate $\pi_s(x)p(\xi'_t|X_T = x)$ as per definitions of ξ_t, ξ'_t and σ' and verify that the right-hand side of Eq. (2.15) is indeed equal to $\pi_t(x)$. □

Before we embark on the dynamics of the signal-based price process, let us compute $p(\xi_t|x)$ in Eq. (2.14). Indeed, Eq. (2.4) implies $\mathbb{E}[\xi_t|x] = \sigma t x$ and $\mathbb{V}[\xi_t|x] = t/\kappa_t$ where κ_t is as above. Hence,

$$p(\xi_t|x) = \frac{1}{\sqrt{t/\kappa_t}\sqrt{2\pi}} e^{-\frac{1}{2}\frac{(\xi_t - \sigma t x)^2}{t/\kappa_t}}. \tag{2.17}$$

We then accommodate Eq. (2.17) into Eq. (2.14) to get $\pi_t(x)$. With some arrangement,

$$\begin{aligned}
\pi_t(x) &= \frac{p(x)\frac{1}{\sqrt{2\pi}\sqrt{t/\kappa_t}} e^{-\frac{1}{2}\frac{(\xi_t-\sigma t x)^2}{t\kappa_t^{-1}}}}{\int_{\mathbb{X}} p(y)\frac{1}{\sqrt{2\pi}\sqrt{t/\kappa_t}} e^{-\frac{1}{2}\frac{(\xi_t-\sigma t y)^2}{t\kappa_t^{-1}}}\mathrm{d}y} \\
&= \frac{p(x)e^{\frac{1}{2}\frac{-\xi_t^2+2\xi_t\sigma t x-\sigma^2 x^2 t^2}{t\kappa_t^{-1}}}}{\int_{\mathbb{X}} p(y)e^{\frac{1}{2}\frac{-\xi_t^2+2\xi_t\sigma t y-\sigma^2 y^2 t^2}{t\kappa_t^{-1}}}\mathrm{d}y} \\
&= \frac{p(x)e^{\kappa_t(\sigma x\xi_t - \frac{1}{2}\sigma^2 x^2 t)}}{\int_{\mathbb{X}} p(y)e^{\kappa_t(\sigma y\xi_t - \frac{1}{2}\sigma^2 y^2 t)}\mathrm{d}y} \quad (x \in \mathbb{X}).
\end{aligned} \tag{2.18}$$

Essentially, Eq. (2.18) is a convolution density in which $p(\xi_t|x)$, as given in Eq. (2.17), operates as a filter on $p(x)$ to map the latter to its posterior $\pi_t(x)$ by comparing the signal ξ_t against each possible value of $x \in \mathbb{X}$.

2.2 The Signal-Based Price Process

We shall continue to assume, w.l.o.g., that the asset pays a single cashflow $\phi(X_T)$ based on a single market factor X_T. Accommodating Eq. (2.18) into Eq. (2.12), the price process $(S_t)_{0\le t\le T}$ can be written as:

$$S_t = \mathbf{1}_{\{t<T\}} e^{-r(T-t)} \frac{\int_{\mathbb{X}} \phi(x)p(x)e^{\kappa_t\left(\sigma x\xi_t - \frac{1}{2}\sigma^2 x^2 t\right)}\mathrm{d}x}{\int_{\mathbb{X}} p(x)e^{\kappa_t\left(\sigma x\xi_t - \frac{1}{2}\sigma^2 x^2 t\right)}\mathrm{d}x}. \tag{2.19}$$

The dynamics of S_t, on the other hand, can be worked out as follows. First, we re-write Eq. (2.11):

$$S_t = \mathbf{1}_{\{t<T\}} e^{-r(T-t)} \mathbb{E}\left[\phi(X_T)|\xi_t\right] = \mathbf{1}_{\{t<T\}} e^{-r(T-t)} \phi_t(X_T), \tag{2.20}$$

where, again, $\phi_t(X_T) = \mathbb{E}\left[\phi(X_T)|\xi_t\right]$. Apparently, $\phi_t(X_T)$ can be expressed in the form $\phi(t, \xi_t)$. Equation (2.19), on the other hand, implies:

$$\phi(t, \xi_t) = \frac{\int_{\mathbb{X}} \phi(x)p(x)e^{\kappa_t\left(\sigma x\xi_t - \frac{1}{2}\sigma^2 x^2 t\right)}\mathrm{d}x}{\int_{\mathbb{X}} p(x)e^{\kappa_t\left(\sigma x\xi_t - \frac{1}{2}\sigma^2 x^2 t\right)}\mathrm{d}x}. \tag{2.21}$$

Now, we work out the dynamics of $\phi(t, \xi_t)$, from which the dynamics of S_t will follow directly. Itô's Lemma implies

$$\mathrm{d}\phi(t, \xi_t) = \frac{\partial\phi(t, \xi_t)}{\partial t}\mathrm{d}t + \frac{\partial\phi(t, \xi_t)}{\partial \xi_t}\mathrm{d}\xi_t, \tag{2.22}$$

where the first partial derivative on the right-hand side equals, by virtue of Eq. (2.21),

$$\begin{aligned}\frac{\partial\phi(t, \xi_t)}{\partial t} = \kappa_t^2 \Bigg[&\int_{\mathbb{X}} \phi(x)\left(\sigma\xi_t x/T - \frac{1}{2}\sigma^2 x^2\right) p(x)e^{\kappa_t\left(\sigma x\xi_t - \frac{1}{2}\sigma^2 x^2 t\right)}\mathrm{d}x \int_{\mathbb{X}} p(x)e^{\kappa_t\left(\sigma x\xi_t - \frac{1}{2}\sigma^2 x^2 t\right)}\mathrm{d}x \\ &- \int_{\mathbb{X}} \phi(x)p(x)e^{\kappa_t\left(\sigma x\xi_t - \frac{1}{2}\sigma^2 x^2 t\right)}\mathrm{d}x \int_{\mathbb{X}} \left(\sigma\xi_t x/T - \frac{1}{2}\sigma^2 x^2\right) p(x)e^{\kappa_t\left(\sigma x\xi_t - \frac{1}{2}\sigma^2 x^2 t\right)}\mathrm{d}x\Bigg]\end{aligned}$$

$$
\begin{aligned}
&\cdot\left(\int_{\mathbb{X}} p(x) e^{\kappa_t\left(\sigma x \xi_t-\frac{1}{2}\sigma^2 x^2 t\right)} \mathrm{d}x\right)^{-2} \\
&=\kappa_t^2\Big[(\sigma\xi_t/T)\,\mathbb{E}_t\left[\phi(X_T)X_T\right]-\frac{1}{2}\sigma^2\mathbb{E}_t\left[\phi(X_T)X_T^2\right] \\
&\quad -\mathbb{E}_t\left[\phi(X_T)\right]\left((\sigma\xi_t/T)\,\mathbb{E}_t\left[X_T\right]-\frac{1}{2}\sigma^2\mathbb{E}_t\left[X_T^2\right]\right)\Big] \\
&=\kappa_t^2\Big[(\sigma\xi_t/T)\left(\mathbb{E}_t\left[\phi(X_T)X_T\right]-\mathbb{E}_t\left[\phi(X_T)\right]\mathbb{E}_t\left[X_T\right]\right) \\
&\quad -\frac{1}{2}\sigma^2\left(\mathbb{E}_t\left[\phi(X_T)X_T^2\right]-\mathbb{E}_t\left[\phi(X_T)\right]\mathbb{E}_t\left[X_T^2\right]\right)\Big] \\
&=\kappa_t^2\left[(\sigma\xi_t/T)\,\mathbb{C}\mathrm{ov}_t\left(\phi(X_T),X_T\right)-\sigma^2\mathbb{E}_t\left[\phi(X_T)\right]\mathbb{C}\mathrm{ov}_t\left(\phi(X_T),X_T\right)\right] \\
&=\sigma\kappa_t^2\left(\xi_t/T-\sigma\mathbb{E}_t\left[\phi(X_T)\right]\right)\mathbb{C}\mathrm{ov}_t\left(\phi(X_T),X_T\right).
\end{aligned}
\tag{2.23}
$$

where $\mathbb{C}\mathrm{ov}_t = \mathbb{C}\mathrm{ov}(\cdot|\xi_t)$ denotes the conditional covariance with respect to available information ξ_t at time t. The equality

$$
\frac{1}{2}\left(\mathbb{E}_t\left[\phi(X_T)X_T^2\right]-\mathbb{E}_t\left[\phi(X_T)\right]\mathbb{E}_t\left[X_T^2\right]\right)=\mathbb{E}_t\left[\phi(X_T)\right]\mathbb{C}\mathrm{ov}_t\left(\phi(X_T),X_T\right) \tag{2.24}
$$

in Eq. (2.23) can indeed be verified by applying Stein's Lemma (cf. [32]), which implies that

$$
\mathbb{E}\left[g(X)\left(X-\mathbb{E}\left[X\right]\right)\right]=\mathbb{V}\left(X\right)\mathbb{E}\left[g'\left(X\right)\right] \tag{2.25}
$$

holds for any differentiable function g. Similarly, one can verify that the second partial derivative term on the right-hand side of Eq. (2.22) is equivalent to

$$
\begin{aligned}
\frac{\partial\phi(t,\xi_t)}{\partial\xi_t}&=\sigma\kappa_t\Big[\int_{\mathbb{X}}\phi(x)xp(x)e^{\kappa_t\left(\sigma x\xi-\frac{1}{2}\sigma^2x^2t\right)}\mathrm{d}x\int_{\mathbb{X}}p(x)e^{\kappa_t\left(\sigma x\xi_t-\frac{1}{2}\sigma^2x^2t\right)}\mathrm{d}x \\
&\quad -\int_{\mathbb{X}}\phi(x)p(x)e^{\kappa_t\left(\sigma x\xi_t-\frac{1}{2}\sigma^2x^2t\right)}\mathrm{d}x\int_{\mathbb{X}}xp(x)e^{\kappa_t\left(\sigma x\xi_t-\frac{1}{2}\sigma^2x^2t\right)}\mathrm{d}x\Big] \\
&\quad \cdot\left(\int_{\mathbb{X}}p(x)e^{\kappa_t\left(\sigma x\xi_t-\frac{1}{2}\sigma^2x^2t\right)}\mathrm{d}x\right)^{-2} \\
&=\sigma\kappa_t\left(\mathbb{E}_t\left[\phi(X_T)X_T\right]-\mathbb{E}_t\left[\phi(X_T)\right]\mathbb{E}_t\left[X_T\right]\right) \\
&=\sigma\kappa_t\mathbb{C}\mathrm{ov}_t\left(\phi(X_T),X_T\right).
\end{aligned}
\tag{2.26}
$$

Combining Eqs. (2.23) and (2.26) as per Eq. (2.22) yields

$$\mathrm{d}\phi_t(X_T) = \sigma\kappa_t\mathbb{C}\mathrm{ov}_t\left(\phi(X_T), X_T\right)\left[\kappa_t\left(\xi_t/T - \sigma\mathbb{E}_t\left[\phi(X_T)\right]\right)\mathrm{d}t + \mathrm{d}\xi_t\right]. \tag{2.27}$$

As a direct result of Eq. (2.27), and using $S_t = e^{-r(T-t)}\phi_t(X_T)$, the dynamics of S_t is given by

$$\begin{aligned}\mathrm{d}S_t &= re^{-r(T-t)}\phi_t(X_T)\mathrm{d}t + e^{-r(T-t)}\mathrm{d}\phi_t(X_T),\\ &= re^{-r(T-t)}\phi_t(X_T)\mathrm{d}t + e^{-r(T-t)}\sigma\kappa_t\mathbb{C}\mathrm{ov}_t\left(\phi(X_T), X_T\right)\\ &\quad\cdot\left[\kappa_t\left(\xi_t/T - \sigma\mathbb{E}_t\left[\phi(X_T)\right]\right)\mathrm{d}t + \mathrm{d}\xi_t\right],\\ &= rS_t\mathrm{d}t + \Lambda_t\mathrm{d}W_t,\end{aligned} \tag{2.28}$$

where $\Lambda_t := e^{-r(T-t)}\sigma\kappa_t\mathbb{C}\mathrm{ov}_t\left(\phi(X_T), X_T\right)$, and

$$\mathrm{d}W_t := \kappa_t\left(\xi_t/T - \sigma\phi_t(X_T)\right)\mathrm{d}t + \mathrm{d}\xi_t \tag{2.29}$$

or, equivalently,

$$W_t := \xi_t + \int_0^t \kappa_s\left(\xi_s/T - \sigma\phi_s(X_T)\right)\mathrm{d}s. \tag{2.30}$$

Alternatively, one can start with the dynamics of $\pi_t(x)$ given in Eq. (2.18) and use Eq. (2.12) to reach the same result as in Eq. (2.28) (for details, see [10, 24]).

The term Λ_t which appears in Eq. (2.28) is the 'absolute price volatility.'[5] An interesting observation related to the absolute price volatility is that its overall magnitude is determined by the signal-to-noise parameter σ. Thus, an increase in the information flow rate means an increased price volatility. This observation seems to be paradoxical if one considers that the growing financial market interconnectedness—which is expected to increase market efficiency and reduce price anomalies—can actually increase the price volatility.

Another interesting observation from Eq. (2.28) pertains to W_t and helps us shed a bit more light on another somewhat paradoxical point in financial mathematics literature which pertains to whether W_t really contains information or represents pure noise. The process W_t, as will be shown to be a martingale w.r.t. F_t^{ξ} below, is not imposed on the model as one of its inputs, but rather appeared as one of its by-products. Furthermore, when the information flow structure is defined explicitly, W_t is no more simply irreducible, as suggested by many classical models in financial mathematics.

Proposition 2.3 *The process W_t in Eq.* (2.30) *is a Brownian motion adapted to F_t^{ξ}.*

[5]Note that Λ_t is also forward-looking.

Proof (See [10, 24] for a Sketch) Referring to Lévy's characterisation of F_t-Brownian motion, we first need to show that the process W_t, as defined in Eq. (2.29) is an F_t^{ξ}-martingale and, second, that $\mathrm{d}\,[W, W]\,(t) = \mathrm{d}t$. The first condition is equivalent to

$$\mathbb{E}\left[W_u | \mathcal{F}_t^{\xi}\right] = W_t \quad \text{or} \quad \mathbb{E}\left[W_u - W_t | \mathcal{F}_t^{\xi}\right] = 0 \tag{2.31}$$

for $u \geq t$. Using the definition of W_t in Eq. (2.29) as well as the Markov property of the process ξ_t, we can write the left-hand side more explicitly as

$$\begin{aligned}\mathbb{E}\left[W_u - W_t | \mathcal{F}_t^{\xi}\right] &= \mathbb{E}\left[\xi_u - \xi_t | \xi_t\right] + \mathbb{E}\left[\int_t^u (\kappa_s \xi_s / T)\,\mathrm{d}s | \xi_t\right] \\ &\quad - \mathbb{E}\left[\int_t^u \sigma \kappa_s \mathbb{E}\left[\phi(X_T) | \xi_s\right] \mathrm{d}s | \xi_t\right].\end{aligned} \tag{2.32}$$

Now, using the definition of ξ_t in Eq. (2.4) in the first and second terms on the right-hand side, and the tower property[6] in the third one, we find

$$\begin{aligned}\mathbb{E}\left[W_u - W_t | \mathcal{F}_t^{\xi}\right] &= \sigma u \mathbb{E}\left[\phi(X_T) | \xi_t\right] + \mathbb{E}\left[\beta_u | \xi_t\right] - \left(\sigma t \mathbb{E}\left[\phi(X_T) | \xi_t\right] + \mathbb{E}\left[\beta_t | \xi_t\right]\right) \\ &\quad + \mathbb{E}\left[\int_t^u (\kappa_s / T)\left(\sigma s \phi(X_T) + \beta_s\right) ds | \xi_t\right] \\ &\quad - \sigma \mathbb{E}\left[\phi(X_T) | \xi_t\right] \int_t^u \kappa_s \mathrm{d}s \\ &= \sigma u \mathbb{E}\left[\phi(X_T) | \xi_t\right] + \mathbb{E}\left[\beta_u | \xi_t\right] - \left(\sigma t \mathbb{E}\left[\phi(X_T) | \xi_t\right] + \mathbb{E}\left[\beta_t | \xi_t\right]\right) \\ &\quad + (\sigma / T)\,\mathbb{E}\left[\phi(X_T) | \xi_t\right] \int_t^u s \kappa_s \mathrm{d}s + (1/T)\,\mathbb{E}\left[\int_t^u \kappa_s \beta_s \mathrm{d}s | \xi_t\right] \\ &\quad - \sigma \mathbb{E}\left[\phi(X_T) | \xi_t\right] \int_t^u \kappa_s \mathrm{d}s.\end{aligned} \tag{2.33}$$

The integral $(1/T) \displaystyle\int_t^u s \kappa_s \mathrm{d}s$ can be shown to be equal to $\displaystyle\int_t^u \kappa_s \mathrm{d}s + t - u$ by change of variable $T - s$ to ν, and, then, ν back to $T - s$. Hence,

$$\begin{aligned}\mathbb{E}\left[W_u - W_t | \mathcal{F}_t^{\xi}\right] &= \sigma u \mathbb{E}\left[\phi(X_T) | \xi_t\right] + \mathbb{E}\left[\beta_u | \xi_t\right] - \left(\sigma t \mathbb{E}\left[\phi(X_T) | \xi_t\right] + \mathbb{E}\left[\beta_t | \xi_t\right]\right) \\ &\quad + \sigma \mathbb{E}\left[\phi(X_T) | \xi_t\right] \left(\int_t^u \kappa_s \mathrm{d}s + t - u)\right) + (1/T)\,\mathbb{E}\left[\int_t^u \kappa_s \beta_s \mathrm{d}s | \xi_t\right]\end{aligned}$$

[6] $\mathbb{E}\left[\mathbb{E}\left[X | \mathcal{F}_u\right] | \mathcal{F}_t\right] = \mathbb{E}\left[X | \mathcal{F}_t\right]$ for $t \leq u$ and increasing set of σ-algebras $(\mathcal{F}_t)_{t \geq 0}$.

$$-\sigma \mathbb{E}\left[\phi(X_T)|\xi_t\right] \int_t^u \kappa_s \mathrm{d}s$$

$$= \mathbb{E}\left[\beta_u|\xi_t\right] - \mathbb{E}\left[\beta_t|\xi_t\right] + (1/T)\,\mathbb{E}\left[\int_t^u \kappa_s \beta_s \mathrm{d}s|\xi_t\right]. \tag{2.34}$$

To conclude that W_t is an F_t^{ξ}-martingale, we write the expectation $\mathbb{E}\left[\beta_u|\xi_t\right]$ above as follows:

$$\mathbb{E}\left[\beta_u|\xi_t\right] = \mathbb{E}\left[\mathbb{E}\left[\beta_u|\phi(X_T), \beta_t\right]|\xi_t\right]. \tag{2.35}$$

Note, by intuition, that the σ-algebra generated by both $\phi(X_T)$ and β_t is larger than that by ξ_t (the agent, given ξ_t, cannot know what is noise and what is not). Since we also know that the independence relation $\phi(X_T) \perp\!\!\!\perp \beta_u$ exists, Eq. (2.35) can be rearranged as

$$\mathbb{E}\left[\beta_u|\xi_t\right] = \mathbb{E}\left[\mathbb{E}\left[\beta_u|\beta_t\right]|\xi_t\right]. \tag{2.36}$$

To calculate $\mathbb{E}\left[\beta_u|\beta_t\right]$, we use the independence relation[7]

$$\beta_u \kappa_u - \beta_t \kappa_t \perp\!\!\!\perp \beta_t. \tag{2.37}$$

Also, we note that $\mathbb{E}\left[\beta_u|\beta_t\right]$ can be rearranged as

$$\begin{aligned}\mathbb{E}\left[\beta_u|\beta_t\right] &= \kappa_u^{-1}\left(\mathbb{E}\left[\beta_u\kappa_u - \beta_t\kappa_t|\beta_t\right] + \mathbb{E}\left[\beta_t\kappa_t|\beta_t\right]\right)\\ &= \kappa_u^{-1}\left(\mathbb{E}\left[\beta_u\kappa_u - \beta_t\kappa_t|\beta_t\right] + \beta_t\kappa_t\right).\end{aligned} \tag{2.38}$$

Using the independence relation in (2.37) then implies $\mathbb{E}\left[\beta_u|\beta_t\right] = (\kappa_t/\kappa_u)\beta_t$ and, therefore, $\mathbb{E}\left[\mathbb{E}\left[\beta_{uT}|\beta_{tT}\right]|\xi_t\right] = (\kappa_t/\kappa_u)\mathbb{E}\left[\beta_{tT}|\xi_t\right]$. Then, we conclude that W_t is an F_t^{ξ}-martingale by employing Eq. (2.38) in Eq. (2.34), i.e.,

$$\begin{aligned}\mathbb{E}\left[W_u - W_t|\mathcal{F}_t^{\xi}\right] &= (\kappa_t/\kappa_u - 1)\mathbb{E}\left[\beta_{tT}|\xi_t\right] + (1/T)\int_t^u \kappa_s(\kappa_t/\kappa_s)\mathbb{E}\left[\beta_t|\xi_t\right]\mathrm{d}s\\ &= (\kappa_t/\kappa_u - 1)\mathbb{E}\left[\beta_{tT}|\xi_t\right] + ((u-t)\kappa_t/T)\,\mathbb{E}\left[\beta_t|\xi_t\right]\\ &= (\kappa_t/\kappa_u - 1)\mathbb{E}\left[\beta_{tT}|\xi_t\right] + (1 - \kappa_t/\kappa_u)\mathbb{E}\left[\beta_t|\xi_t\right]\\ &= 0.\end{aligned} \tag{2.39}$$

The second part, i.e., $\mathrm{d}\left[W, W\right](t) = \mathrm{d}t$, is rather simple. Recall Eq. (2.29), i.e.,

$$\mathrm{d}W_t := \kappa_t\left(\xi_t/T - \sigma\mathbb{E}_t\left[\phi(X_T)\right]\right)\mathrm{d}t + \mathrm{d}\xi_t. \tag{2.40}$$

[7]Indeed; $\mathbb{V}\left[\beta_u\kappa_u - \beta_t\kappa_t, \beta_t\right] = \mathbb{E}\left[(\beta_u\kappa_u - \beta_t\kappa_t)\beta_t\right] = \kappa_u\mathbb{E}\left[\beta_u\beta_t\right] - \kappa_t\mathbb{E}\left[\beta_t^2\right] = t\kappa_u\kappa_u^{-1} - t\kappa_t\kappa_t^{-1} = 0$.

We note that $\mathrm{d}[W,W](t)$ is only due to $\mathrm{d}[\xi,\xi](t)$ (given ξ_t at time t). And, by representation (2.9), we already know that

$$\mathrm{d}[\xi,\xi](t) = \mathrm{d}[B,B](t) = \mathrm{d}t, \tag{2.41}$$

which completes the proof. □

Below, we will work out S_t for some particular dividend structures.

2.2.1 Gaussian Dividends

Assume ϕ is an identity, i.e., $\phi(X_T) = X_T$ with $X_T \sim \mathcal{N}(0,1)$. Then, accommodating two well-known Gaussian integrals[8] into Eq. (2.19) would yield

$$\begin{aligned}
S_t &= \mathbf{1}_{\{t<T\}} e^{-r(T-t)} \frac{\int_{\mathbb{X}} x p(x) e^{\kappa_t\left(\sigma x \xi_t - \frac{1}{2}\sigma^2 x^2 t\right)} \mathrm{d}x}{\int_{\mathbb{X}} p(x) e^{\kappa_t\left(\sigma x \xi_t - \frac{1}{2}\sigma^2 x^2 t\right)} \mathrm{d}x} \\
&= \mathbf{1}_{\{t<T\}} e^{-r(T-t)} \frac{\int_{\mathbb{X}} x e^{-\frac{x^2}{2}} e^{(\sigma\kappa_t\xi_t)x - \left(\frac{1}{2}\sigma^2\kappa_t t\right)x^2} \mathrm{d}x}{\int_{\mathbb{X}} e^{-\frac{x^2}{2}} e^{(\sigma\kappa_t\xi_t)x - \left(\frac{1}{2}\sigma^2\kappa_t t\right)x^2} \mathrm{d}x} \\
&= \mathbf{1}_{\{t<T\}} e^{-r(T-t)} \frac{\sigma\kappa_t\xi_t \left(\sigma^2\kappa_t t + 1\right)^{-3/2}}{\left(\sigma^2\kappa_t t + 1\right)^{-1/2}} \\
&= \mathbf{1}_{\{t<T\}} e^{-r(T-t)} \frac{\sigma\kappa_t\xi_t}{\sigma^2\kappa_t t + 1},
\end{aligned} \tag{2.42}$$

where $\mathbb{X} = (-\infty, \infty)$. Since κ_t has a singularity at $t = T$ (i.e., $\kappa_t \to \infty$ as $t \to T$), we can talk about the limit of S_t as t approaches T. Indeed, it is straightforward to show that

$$\lim_{(T-t)\to 0} S_t = \frac{\xi_t}{\sigma T} = X_T. \tag{2.43}$$

[8]These are [27]:

1. $\int_{\mathbb{X}} \exp(-x^2/2)\exp(ax - bx^2)\mathrm{d}x = \sqrt{2\pi}(2b+1)^{-1/2}\exp(a^2/(2(2b+1)))$, and
2. $\int_{\mathbb{X}} x\exp(-x^2/2)\exp(ax - bx^2)\mathrm{d}x = \sqrt{2\pi}a(2b+1)^{-3/2}\exp(a^2/(2(2b+1)))$,

where $\mathbb{X} = (-\infty, \infty)$.

2.2.2 Exponential Dividends

Besides normal density, exponential class of distributions are also commonly used to model dividends. Assume, again, ϕ is an identity function and $p(\phi(x)) = p(x) = \lambda e^{-\lambda x}$, $\lambda > 0$, and $\mathbb{X} = [0, \infty)$, a priori. This implies, again by virtue of Eq. (2.19), that

$$
\begin{aligned}
S_t &= \mathbf{1}_{\{t<T\}} e^{-r(T-t)} \frac{\displaystyle\int_{\mathbb{X}} x e^{-\lambda x} e^{\kappa_t\left(\sigma x \xi_t - \frac{1}{2}\sigma^2 x^2 t\right)} \mathrm{d}x}{\displaystyle\int_{\mathbb{X}} e^{-\lambda x} e^{\kappa_t\left(\sigma x \xi_t - \frac{1}{2}\sigma^2 x^2 t\right)} \mathrm{d}x} \\
&= \mathbf{1}_{\{t<T\}} e^{-r(T-t)} \frac{\displaystyle\int_{\mathbb{X}} x e^{-\frac{1}{2}\left(\left(x\sigma\sqrt{t\kappa_t}\right)^2 - 2x(\kappa_t\sigma\xi_t - \lambda) + a^2\right)} \mathrm{d}x}{\displaystyle\int_{\mathbb{X}} e^{-\frac{1}{2}\left(\left(x\sigma\sqrt{t\kappa_t}\right)^2 - 2x(\kappa_t\sigma\xi_t - \lambda) + a^2\right)} \mathrm{d}x} \\
&= \mathbf{1}_{\{t<T\}} e^{-r(T-t)} \frac{\displaystyle\int_{\mathbb{X}} x e^{-\frac{1}{2}\left(x\sigma\sqrt{t\kappa_t} - a\right)^2} \mathrm{d}x}{\displaystyle\int_{\mathbb{X}} e^{-\frac{1}{2}\left(x\sigma\sqrt{t\kappa_t} - a\right)^2} \mathrm{d}x} \qquad \left(a = \frac{\kappa_t\sigma\xi_t - \lambda}{\sigma\sqrt{t\kappa_t}}\right) \\
&= \mathbf{1}_{\{t<T\}} e^{-r(T-t)} \frac{\displaystyle\int_{\mathbb{X}'} \frac{x' + a}{\sigma\sqrt{t\kappa_t}} e^{-x'^2/2} \mathrm{d}x'}{\displaystyle\int_{\mathbb{X}'} e^{-x'^2/2} \mathrm{d}x'} \qquad \left(x' = x\sigma\sqrt{t\kappa_t} - a, \quad \mathbb{X}' = [-a, \infty)\right)
\end{aligned}
\tag{2.44}
$$

With some further arrangement, we obtain the following explicit formula for the asset price (similar to the one in [10]):

$$
\begin{aligned}
S_t &= \mathbf{1}_{\{t<T\}} e^{-r(T-t)} \frac{\frac{1}{\sigma\sqrt{t\kappa_t}} \displaystyle\int_{\mathbb{X}'} x' e^{-x'^2/2} \mathrm{d}x' + \frac{a}{\sigma\sqrt{t\kappa_t}} \sqrt{2\pi}\,(1 - \Theta(-a))}{\sqrt{2\pi}\,(1 - \Theta(-a))} \qquad \left(y = x'^2, \quad \mathrm{d}y = 2x'\mathrm{d}x'\right) \\
&= \mathbf{1}_{\{t<T\}} e^{-r(T-t)} \frac{\frac{1}{\sigma\sqrt{t\kappa_t}} \left(-e^{-x'^2/2}\big|_{\mathbb{X}'}\right) + \frac{a}{\sigma\sqrt{t\kappa_t}} \sqrt{2\pi}\,\Theta(a)}{\sqrt{2\pi}\,\Theta(a)} \\
&= \mathbf{1}_{\{t<T\}} e^{-r(T-t)} \frac{1}{\sigma\sqrt{t\kappa_t}} \left(\frac{e^{-\frac{1}{2}\left(\frac{\kappa_t\sigma\xi_t - \lambda}{\sigma\sqrt{t\kappa_t}}\right)^2}}{\sqrt{2\pi}\,\Theta\left(\frac{\kappa_t\sigma\xi_t - \lambda}{\sigma\sqrt{t\kappa_t}}\right)} + \frac{\kappa_t\sigma\xi_t - \lambda}{\sigma\sqrt{t\kappa_t}} \right).
\end{aligned}
\tag{2.45}
$$

where $\Theta(\cdot)$ denotes the standard normal cumulative density.

2.2.3 Log-Normal Dividends

Assume now $\phi(X_T)$ is not an identity but the dividend $\phi(X_T)$ will be paid according to

$$\phi(X_T) = S_0 e^{(\mu - \frac{1}{2}\nu^2)T + \nu\sqrt{T}X_T}, \tag{2.46}$$

with $\nu > 0$, and let $X_T \sim \mathcal{N}(0,1)$. Note that this is equivalent to saying that $\phi(X_T)$ is an identity and adheres to lognormal marginal law with parameters $(\ln S_0 + (\mu - \nu^2/2)T, \nu\sqrt{T})$. We now simply accommodate Eq. (2.46) into Eq. (2.19), i.e.,

$$\begin{aligned}
S_t &= \mathbf{1}_{\{t<T\}} e^{-r(T-t)} \frac{\int_{\mathbb{X}} S_0 e^{(\mu-\frac{1}{2}\nu^2)T+\nu\sqrt{T}x} e^{-\frac{x^2}{2}} e^{\kappa_t(\sigma x \xi_t - \frac{1}{2}\sigma^2 x^2 t)} \mathrm{d}x}{\int_{\mathbb{X}} e^{-\frac{x^2}{2}} e^{\kappa_t(\sigma x \xi_t - \frac{1}{2}\sigma^2 x^2 t)} \mathrm{d}x} \\
&= \mathbf{1}_{\{t<T\}} e^{-r(T-t)} S_0 e^{(\mu-\frac{1}{2}\nu^2)T} \frac{\int_{\mathbb{X}} e^{\nu\sqrt{T}x} e^{-\frac{x^2}{2}} e^{\kappa_t(\sigma x \xi_t - \frac{1}{2}\sigma^2 x^2 t)} \mathrm{d}x}{\int_{\mathbb{X}} e^{-\frac{x^2}{2}} e^{\kappa_t(\sigma x \xi_t - \frac{1}{2}\sigma^2 x^2 t)} \mathrm{d}x} \\
&= \mathbf{1}_{\{t<T\}} e^{-r(T-t)} S_0 e^{(\mu-\frac{1}{2}\nu^2)T} \frac{e^{\frac{1}{2}a^2}}{e^{\frac{1}{2}b^2}} \frac{\int_{\mathbb{X}} e^{-\frac{1}{2}(x^2(1+\sigma^2\kappa_t t) - 2x(\nu\sqrt{T}+\kappa_t\sigma\xi_t) + a^2)} \mathrm{d}x}{\int_{\mathbb{X}} e^{-\frac{1}{2}(x^2(1+\sigma^2\kappa_t t) - 2\kappa_t\sigma\xi_t + b^2)} \mathrm{d}x} \\
&= \mathbf{1}_{\{t<T\}} e^{-r(T-t)} S_0 e^{(\mu-\frac{1}{2}\nu^2)T} \frac{e^{\frac{1}{2}a^2}}{e^{\frac{1}{2}b^2}} \frac{\int_{\mathbb{X}} e^{-\frac{1}{2}\left(x\sqrt{1+\sigma^2\kappa_t t} - a\right)^2} \mathrm{d}x}{\int_{\mathbb{X}} e^{-\frac{1}{2}\left(x\sqrt{1+\sigma^2\kappa_t t} - b\right)^2} \mathrm{d}x},
\end{aligned} \tag{2.47}$$

where $\mathbb{X} = (-\infty, \infty)$, $a = (\nu\sqrt{T} + \kappa_t\sigma\xi_t)/\sqrt{1+\sigma^2\kappa_t t}$ and $b = (\kappa_t\sigma\xi_t)/\sqrt{1+\sigma^2\kappa_t t}$. Hence, we get

$$S_t = \mathbf{1}_{\{t<T\}} e^{-r(T-t)} S_0 e^{\left(\mu - \frac{\nu^2}{2}\right)T + \frac{\nu^2 T}{2(1+\sigma^2\kappa_t t)} + \frac{\nu\sqrt{T}\kappa_t\sigma}{1+\sigma^2\kappa_t t}\xi_t}. \tag{2.48}$$

As for dynamics $\mathrm{d}S_t$, we first need the evaluate the conditional covariance term which appears in Eq. (2.27), i.e,

$$\mathrm{d}\phi_t(X_T) = \sigma\kappa_t \mathbb{C}\mathrm{ov}_t\left[\phi(X_T), X_T\right]\left[\kappa_t\left(T^{-1}\xi_t - \sigma\mathbb{E}_t\left[\phi(X_T)\right]\right)\mathrm{d}t + \mathrm{d}\xi_t\right], \tag{2.49}$$

to derive the dynamics of S_t as given in Eq. (2.28), i.e.,

$$\mathrm{d}S_t = rS_t\mathrm{d}t + \Lambda_t\mathrm{d}W_t, \tag{2.50}$$

where, again, $\Lambda_t := e^{-r(T-t)}\sigma\kappa_t\mathbb{C}\mathrm{ov}_t\left(\phi(X_T), X_T\right)$, and $\mathrm{d}W_t$ and W_t as in Eqs. (2.29) and (2.30), respectively.

In order to calculate the conditional covariance term and, hence, $\mathrm{d}S_t$, more explicitly, we need to evaluate one additional integral, namely,

$$\mathbb{E}_t\left[\phi(X_T)X_T\right] = \frac{\displaystyle\int_{\mathbb{X}} \phi(x)xe^{-\frac{x^2}{2}}e^{\kappa_t\left(\sigma x\xi_t - \frac{1}{2}\sigma^2x^2t\right)}\mathrm{d}x}{\displaystyle\int_{\mathbb{X}} e^{-\frac{x^2}{2}}e^{\kappa_t\left(\sigma x\xi_t - \frac{1}{2}\sigma^2x^2t\right)}\mathrm{d}x}. \tag{2.51}$$

We can use the same Gaussian integrals as above to compute Eq. (2.51):

$$\begin{aligned}
\mathbb{E}_t\left[\phi(X_T)X_T\right] &= \frac{S_0e^{\left(\mu-\frac{1}{2}\nu^2\right)T}\displaystyle\int_{\mathbb{X}} xe^{-\frac{x^2}{2}}e^{\left(\kappa_t\sigma\xi_t+\nu\sqrt{T}\right)x-\left(\frac{1}{2}\kappa_t\sigma^2t\right)x^2}\mathrm{d}x}{\displaystyle\int_{\mathbb{X}} e^{-\frac{x^2}{2}}e^{\left(\kappa_t\sigma\xi_t\right)x-\left(\frac{1}{2}\kappa_t\sigma^2t\right)x^2}\mathrm{d}x} \\
&= \frac{S_0e^{\left(\mu-\frac{1}{2}\nu^2\right)T}\frac{\kappa_t\sigma\xi_t+\nu\sqrt{T}}{\left(\kappa_t\sigma^2t+1\right)^{3/2}}\exp\left(\frac{1}{2}\frac{\left(\kappa_t\sigma\xi_t+\nu\sqrt{T}\right)^2}{\sigma^2\kappa_tt+1}\right)}{\frac{1}{\left(\sigma^2\kappa_tt+1\right)^{1/2}}\exp\left(\frac{1}{2}\frac{\left(\sigma\kappa_t\xi_t\right)^2}{\sigma^2\kappa_tt+1}\right)} \\
&= S_0e^{\left(\mu-\frac{1}{2}\nu^2\right)T}\frac{\sigma\kappa_t\xi_t+\nu\sqrt{T}}{\sigma^2\kappa_tt+1}\exp\left(\frac{\sigma\kappa_t\xi_t\nu\sqrt{T}+\frac{1}{2}\nu^2T}{\sigma^2\kappa_tt+1}\right).
\end{aligned} \tag{2.52}$$

We are now equipped with Eqs. (2.42), (2.48) and (2.52) to calculate the conditional covariance $\mathbb{C}\mathrm{ov}_t\left(\phi(X_T), X_T\right) = \mathbb{E}_t\left[\phi(X_T)X_T\right] - \mathbb{E}_t\left[\phi(X_T)\right]\mathbb{E}_t\left[X_T\right]$:

$$\begin{aligned}
\mathbb{C}\mathrm{ov}_t\left(\phi(X_T), X_T\right) &= S_0e^{\left(\mu-\frac{1}{2}\nu^2\right)T}\frac{\sigma\kappa_t\xi_t+\nu\sqrt{T}}{\sigma^2\kappa_tt+1}\exp\left(\frac{\sigma\kappa_t\xi_t\nu\sqrt{T}+\frac{1}{2}\nu^2T}{1+\sigma^2\kappa_tt}\right) \\
&\quad -S_0e^{\left(\mu-\frac{1}{2}\nu^2\right)T}\frac{\sigma\kappa_t\xi_t}{\sigma^2\kappa_tt+1}\exp\left(\frac{\frac{1}{2}\nu^2T}{1+\sigma^2\kappa_tt}+\frac{\sigma\kappa_t\xi_t\nu\sqrt{T}}{1+\sigma^2\kappa_tt}\right) \\
&= S_0e^{\left(\mu-\frac{1}{2}\nu^2\right)T}\exp\left(\frac{\sigma\kappa_t\xi_t\nu\sqrt{T}+\frac{1}{2}\nu^2T}{\sigma^2\kappa_tt+1}\right)\frac{\nu\sqrt{T}}{1+\sigma^2\kappa_tt}.
\end{aligned} \tag{2.53}$$

Notice that Eq. (2.53) without its last term is exactly $S_t e^{r(T-t)}$ and, therefore,

$$\mathbb{C}\mathrm{ov}_t\left(\phi(X_T), X_T\right) = S_t e^{r(T-t)} \frac{\nu\sqrt{T}}{1+\sigma^2\kappa_t t}. \tag{2.54}$$

Then, by virtue of Eq. (2.50), we have

$$\mathrm{d}S_t = rS_t\mathrm{d}t + \frac{\sigma\kappa_t\nu\sqrt{T}}{\sigma^2\kappa_t t+1} S_t \mathrm{d}W_t, \tag{2.55}$$

where, again, $\mathrm{d}W_t$ is defined as in Eq. (2.41). Now, we discover that, for the particular choice of $\sigma^2 = 1/T$ in Eq. (2.55), not only we get

$$\mathrm{d}S_t = rS_t\mathrm{d}t + \nu S_t\mathrm{d}W_t, \tag{2.56}$$

but also the information flow process ξ_t turns into a standard Brownian motion, i.e.,

$$\begin{aligned}
\mathbb{E}\left[\xi_s\xi_t\right] &= \mathbb{E}\left[\left(sX_T/\sqrt{T}+\beta_{sT}\right)\left(tX_T/\sqrt{T}+\beta_{tT}\right)\right] \\
&= (st/T)\mathbb{E}\left[X_T^2\right] + (s/\sqrt{T})\mathbb{E}\left[X_T\beta_{tT}\right] \\
&\quad +(t/\sqrt{T})\mathbb{E}\left[X_T\beta_{sT}\right] + \mathbb{E}\left[\beta_{sT}\beta_{tT}\right] \quad (X \perp \beta) \\
&= s - \kappa_t^{-1}s + \kappa_t^{-1}s = s.
\end{aligned} \tag{2.57}$$

Yet, there is a more direct way to see this. Once we choose $\sigma^2 = 1/T$, Eq. (2.48) reduces to

$$S_t = \mathbf{1}_{\{t<T\}} S_0 \exp\left(\left(\mu - \nu^2/2\right)t + \nu\xi_t\right), \tag{2.58}$$

where, again, ξ_t substitutes for W_t, the innovation process. Therefore, with the special choice of $\phi(X_T)$ in Eq. (2.46), we actually end up in a Black-Scholes-type model of stock price dynamics. Next, we will deal with signal-based derivatives pricing.

2.3 Change of Measure and Signal-Based Derivative Pricing

In this section, we show how the present signal-based framework can be used to price derivatives. A standard European call option that is written at time 0 on an asset which is characterised by the price process in Eq. (2.48).[9] Assume that the option

[9] It reads $S_t = \mathbf{1}_{\{t<T\}} S_0 \exp\left(-\mu(T-t) + \left(\mu - \frac{\nu^2}{2}\right)T + \frac{\nu^2 T}{2(1+\sigma^2\kappa_t t)} + \frac{\nu\sqrt{T}\kappa_t\sigma}{1+\sigma^2\kappa_t t}\xi_t\right)$.

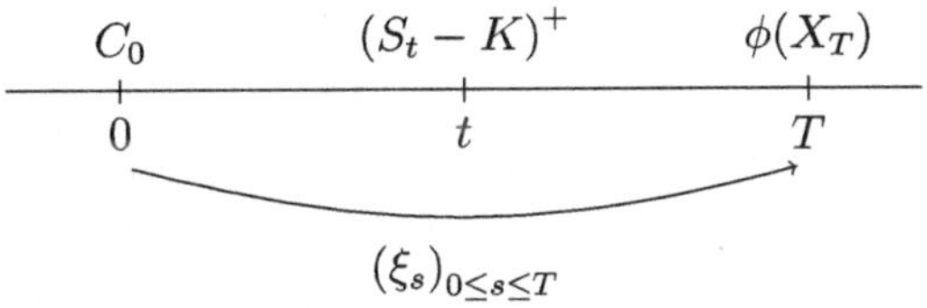

Fig. 2.1 Signal-based option pricing timeline

expires at time t and has an exercise price K. The underlying pays $\phi(X_T) \in (0, \infty)$ at time T where ϕ is not an identity. The information process $(\xi_s)_{0\leq s\leq T}$ will again be carrying signals regarding the factor X_T. Then, the value function for the option can be written as,

$$\begin{aligned} C_0 &= e^{-rt}\mathbb{E}\left[S_t - K\right]^+ \\ &= e^{-rt}\mathbb{E}\left[\phi_t(X_T) - K\right]^+, \end{aligned} \tag{2.59}$$

where $\mathbb{E}[\cdot]^+$ denotes expectation over positive values. A simple timeline is given in Fig. 2.1.

Indeed, Eq. (2.59) looks like an "information" analogous to a forward "rate" agreement (i.e., FIA versus FRA) as, once integrated over all possible values of $\phi(X_T)$, parties in fact contract solely on the time-t value of the pricing signal, i.e., $\{\xi_s\}_{0\leq s\leq T}$, which will determine $\phi(X_T)$. Using Eq. (2.12),[10] we have

$$\begin{aligned} C_0 &= e^{-rt}\mathbb{E}\left[e^{-r(T-t)}\int_{\mathbb{X}}\phi(x)\pi_t(x)\mathrm{d}x - K\right]^+ \\ &= e^{-rt}\mathbb{E}\left[e^{-r(T-t)}\frac{\int_{\mathbb{X}}\phi(x)p(x)e^{\kappa_t\left(\sigma x\xi_t - \frac{1}{2}\sigma^2x^2t\right)}\mathrm{d}x}{\int_{\mathbb{X}}p(x)e^{\kappa_t\left(\sigma x\xi_t - \frac{1}{2}\sigma^2x^2t\right)}\mathrm{d}x} - K\right]^+ \\ &= e^{-rt}\mathbb{E}\left[e^{-r(T-t)}\frac{\int_{\mathbb{X}}\phi(x)p_t(x)\mathrm{d}x}{\int_{\mathbb{X}}p_t(x)\mathrm{d}x} - K\right]^+ \\ &= e^{-rt}\mathbb{E}\left[\left(\int_{\mathbb{X}}p_t(x)\mathrm{d}x\right)^{-1}\int_{\mathbb{X}}\left(e^{-r(T-t)}\phi(x) - K\right)p_t(x)\mathrm{d}x\right]^+ \\ &= e^{-rt}\mathbb{E}\left[\Phi_t^{-1}\int_{\mathbb{X}}\left(e^{-r(T-t)}\phi(x) - K\right)p_t(x)\mathrm{d}x\right]^+, \end{aligned} \tag{2.60}$$

[10] $S_t = \mathbf{1}_{\{t<T\}}e^{-r(T-t)}\int_0^\infty x\pi_t(x)\mathrm{d}x$.

where

$$p_t(x) := p(x)e^{\kappa_t\left(\sigma x\xi_t - \frac{1}{2}\sigma^2 x^2 t\right)}, \quad \Phi_t^{-1} := \left(\int_0^\infty p_t(x)\mathrm{d}x\right)^{-1}. \tag{2.61}$$

Proposition 2.4 *The process Φ_t^{-1} is the Radon–Nikodym derivative (in Girsanov's sense [17]) of the bridge measure $\mathbb{B}$, under which ξ_t turns out to be a standard Brownian bridge, with respect to the pricing measure $\mathbb{Q}$.*

Proof We begin with the dynamics of $p_t = p(t, \xi_t)$. Apparently, using Eq. (2.29)[11] for $\mathrm{d}\xi_t$ and $\mathrm{d}[\xi_t, \xi_t] = \mathrm{d}t$, we have

$$\begin{aligned}
\mathrm{d}p_t &= \frac{\partial p(t,\xi_t)}{\partial t}\mathrm{d}t + \frac{\partial p(t,\xi_t)}{\partial \xi_t}\mathrm{d}\xi_t + \frac{1}{2}\frac{\partial^2 p(t,\xi_t)}{\partial \xi_t^2}\mathrm{d}[\xi_t,\xi_t] \\
&= \left[\kappa_t^2/T\left(\sigma x\xi_t - \frac{1}{2}\sigma^2x^2t\right) - \kappa_t\left(\frac{1}{2}\sigma^2x^2\right)\right]p_t\mathrm{d}t \\
&\quad + \kappa_t\sigma x p_t \mathrm{d}\xi_t + \frac{1}{2}\kappa_t^2\sigma^2x^2p_t\mathrm{d}[\xi_t,\xi_t] \\
&= \left[\kappa_t^2/T\left(\sigma x\xi_t - \frac{1}{2}\sigma^2x^2t\right) - \kappa_t\left(\frac{1}{2}\sigma^2x^2\right)\right]p_t\mathrm{d}t \\
&\quad + \kappa_t\sigma x p_t\left[\mathrm{d}W_t - \kappa_t\left(\xi_t T^{-1} - \sigma\phi_t(X_T)\right)\mathrm{d}t\right] + \frac{1}{2}\kappa_t^2\sigma^2x^2p_t\mathrm{d}t \\
&= \frac{1}{2}\sigma^2x^2\left(\kappa_t^2 - \kappa_t^2 tT^{-1} - \kappa_t\right)p_t\mathrm{d}t + \kappa_t^2\sigma^2x\phi_t(X_T)p_t\mathrm{d}t + \kappa_t\sigma x p_t\mathrm{d}W_t.
\end{aligned} \tag{2.62}$$

The bracketed term in the last line of Eq. (2.62) can easily be shown to equal 0. Therefore,

$$\frac{\mathrm{d}p_t}{p_t} = \kappa_t^2\sigma^2 x\phi_t(X_T)\mathrm{d}t + \kappa_t\sigma x\mathrm{d}W_t. \tag{2.63}$$

Now, we focus on Φ_t. Since $\Phi_t := \int_0^\infty p_t(x)\mathrm{d}x$ is basically a function of time, we can write

$$\mathrm{d}\Phi_t = \mathrm{d}\left(\int_{\mathbb{X}} p_t(x)\mathrm{d}x\right) = \int_{\mathbb{X}} \mathrm{d}p_t(x)\mathrm{d}x. \tag{2.64}$$

---

[11] It can rewritten as $\mathrm{d}\xi_t = \mathrm{d}W_t - \kappa_t\left(\xi_t T^{-1} - \sigma\mathbb{E}_t[\phi(X_T)]\right)\mathrm{d}t$ where, again, W_t is a martingale under the pricing measure.

Accommodating $\mathrm{d}p_t(x)$ from Eq. (2.63), we have

$$\begin{aligned}\mathrm{d}\Phi_t &= \int_{\mathbb{X}} \left(\kappa_t^2 \sigma^2 x p_t(x) \phi_t(X_T) \mathrm{d}t \mathrm{d}x + \int_{\mathbb{X}} \kappa_t \sigma x p_t(x) \mathrm{d}W_t \right) \mathrm{d}x \\ &= \kappa_t^2 \sigma^2 \phi_t(X_T) \left(\int_{\mathbb{X}} x p_t(x) \mathrm{d}x \right) \mathrm{d}t + \kappa_t \sigma \left(\int_{\mathbb{X}} x p_t(x) \mathrm{d}x \right) \mathrm{d}W_t. \end{aligned} \tag{2.65}$$

On the other hand, it follows from the definition of Φ_t that

$$\frac{\int_{\mathbb{X}} x p_t(x) \mathrm{d}x}{\Phi_t} = \phi_t(X_T)$$

$$\text{i.e.,} \quad \int_{\mathbb{X}} x p_t(x) \mathrm{d}x = \phi_t(X_T) \Phi_t. \tag{2.66}$$

Substituting this back into Eq. (2.65), we have

$$\mathrm{d}\Phi_t = \sigma^2 \kappa_t^2 \phi_t^2(X_T) \Phi_t \mathrm{d}t + \sigma \kappa_t \phi_t(X_T) \Phi_t \mathrm{d}W_t. \tag{2.67}$$

And, as a direct consequence,

$$\mathrm{d}\left[\Phi_t, \Phi_t\right] = \sigma^2 \kappa_t^2 \phi_T^2(X_T) \Phi_t^2 \mathrm{d}t. \tag{2.68}$$

Now, applying Itô Formula to $f(\Phi_t) = 1/\Phi_t$, we can show that

$$\begin{aligned} \mathrm{d}f(\Phi_t) = \mathrm{d}\Phi_t^{-1} &= \frac{\partial f(\Phi_t)}{\partial \Phi_t} \mathrm{d}\Phi_t + \frac{1}{2} \frac{\partial^2 f(\Phi_t)}{\partial \Phi_t^2} \mathrm{d}\left[\Phi_t, \Phi_t\right] \\ &= -\Phi_t^{-2} \sigma^2 \kappa_t^2 \Phi_t \phi_T^2(X_T) \mathrm{d}t - \Phi_t^{-2} \sigma \kappa_t \Phi_t \phi_t(X_T) \mathrm{d}W_t \\ &\quad + \Phi_t^{-3} \sigma^2 \kappa_t^2 \phi_t^2(X_T) \Phi_t^2 \mathrm{d}t \\ &= -\sigma^2 \kappa_t^2 \phi_T^2(X_T) f(\Phi_t) \mathrm{d}t - \sigma \kappa_t f(\Phi_t) \phi_t(X_T) \mathrm{d}W_t \\ &\quad + \sigma^2 \kappa_t^2 \phi_t^2(X_T) f(\Phi_t) \mathrm{d}t \\ &= -\sigma \kappa_t \phi_t(X_T) f(\Phi_t) \mathrm{d}W_t, \end{aligned} \tag{2.69}$$

where the term $\sigma \kappa_t \phi_t(X_T)$ can simply be called as the "market price of risk" (again, in Girsanov's sense). Based on Eq. (2.69), $f(\Phi_t)$ can be written as

$$f(\Phi_t) = \exp\left(-\frac{1}{2} \sigma^2 \int_{\mathbb{X}} \kappa_s^2 \phi_s^2(X_T) \mathrm{d}s - \sigma \int_{\mathbb{X}} \kappa_s \phi_s(X_T) \mathrm{d}W_s \right), \tag{2.70}$$

which is the exponential martingale. As a final step, we can define a Brownian motion $\tilde{W}$ under $\mathbb{B}$, by using Girsanov theorem, as

$$\tilde{W}_t := W_t + \sigma \int_{\mathbb{X}} \kappa_s \phi_s(X_T) \mathrm{d}s, \tag{2.71}$$

and check whether $\mathbb{E}^{\mathbb{B}}\left[\tilde{W}_t\right]$ or $\mathbb{E}\left[\tilde{W}_t \Phi_t^{-1}\right]$ is martingale. Apparently, by virtue of Eqs. (2.69) and (2.71),

$$\begin{aligned} \mathrm{d}\left(\tilde{W}_t f(\Phi_t)\right) &= \tilde{W}_t \mathrm{d}f(\Phi_t) + f(\Phi_t)\mathrm{d}\tilde{W}_t + \mathrm{d}\left[\tilde{W}_t, f(\Phi_t)\right] \\ &= -\kappa_t \sigma \phi_t(X_T)\tilde{W}_t f(\Phi_t)\mathrm{d}W_t + f(\Phi_t)\mathrm{d}W_t + \sigma\kappa_t\phi_t(X_T)f(\Phi_t)\mathrm{d}t \\ &\quad -\kappa_t\sigma\phi_t(X_T)f(\Phi_t)\mathrm{d}t \\ &= \left[1 - \tilde{W}_t\kappa_t\sigma\phi_t(X_T)f(\Phi_t)\right]f(\Phi_t)\mathrm{d}W_t \end{aligned} \tag{2.72}$$

is a martingale. This completes the first part of the proof.

We can verify that ξ_t turns into a standard Brownian bridge under $\mathbb{B}$ as follows. If we write W_t explicitly in Eq. (2.71), using Eq. (2.30), indeed, we see that

$$\begin{aligned} \tilde{W}_t &= \xi_t + \int_{\mathbb{X}} \kappa_s \left(\xi_s T^{-1} - \sigma\phi_s(X_T)\right) \mathrm{d}s + \sigma \int_{\mathbb{X}} \kappa_s \phi_s(X_T)\mathrm{d}s \\ &= \xi_t + (1/T) \int_{\mathbb{X}} \kappa_s \xi_s \mathrm{d}s, \end{aligned} \tag{2.73}$$

which implies

$$\xi_t = \tilde{W}_t - (1/T) \int_{\mathbb{X}} \kappa_s \xi_s \mathrm{d}s \quad \text{or} \quad \mathrm{d}\xi_t = \mathrm{d}\tilde{W}_t - (1/T)\kappa_t\xi_t\mathrm{d}t. \tag{2.74}$$

Note that the differential equation in (2.74) above is the one satisfied by a standard Brownian bridge β_{tT} over the interval $[0, T]$. To see this, consider the representation in Eq. (2.9)[12] without the drift term $\phi(x)$.

□

We now rewrite option value in Eq. (2.60) under $\mathbb{B}$ as

$$C_0 = e^{-rt}\mathbb{E}^{\mathbb{B}}\left[\int_{\mathbb{X}} \left(e^{-r(T-t)}\phi(x) - K\right) p_t(x)\mathrm{d}x\right]^+. \tag{2.75}$$

Assuming that there exists a solution ξ_t^* (following from the monotonicity of $p_t(x)$ in ξ_t as per Eq. (2.61)) to the equality

$$\int_{\mathbb{X}} \left(e^{-r(T-t)}\phi(x) - K\right) p_t(x)\mathrm{d}x = 0 \tag{2.76}$$

[12] It reads $\mathrm{d}\xi_t = \left(\sigma\phi(x) - T^{-1}\int_0^t \kappa_s \mathrm{d}W_s'\right)\mathrm{d}t + \mathrm{d}W_t' = \left(\sigma\phi(x) - T^{-1}\kappa_t\xi_t\right)\mathrm{d}t + \mathrm{d}W_t'$.

for arbitrary t, T and K, and knowing that ξ_t is a standard Brownian bridge (i.e., $\xi_t = z\sqrt{t/\kappa_t}$ with $z \sim \mathcal{N}(0,1)$), we can infer C_0 as follows:

$$\begin{aligned}
C_0 &= e^{-rT}\int_{\mathbb{X}} \phi(x)p(x)\int_{\mathbb{Z}} e^{\kappa_t\left(\sigma x z\sqrt{t\kappa_t^{-1}}-\frac{1}{2}\sigma^2x^2t\right)}\frac{1}{\sqrt{2\pi}}e^{-\frac{z^2}{2}}\,\mathrm{d}z\mathrm{d}x \\
&\quad -e^{-rt}K\int_{\mathbb{X}} p(x)\int_{\mathbb{Z}} e^{\kappa_t\left(\sigma x z\sqrt{t\kappa_t^{-1}}-\frac{1}{2}\sigma^2x^2t\right)}\frac{1}{\sqrt{2\pi}}e^{-\frac{z^2}{2}}\,\mathrm{d}z\mathrm{d}x \\
&= e^{-rT}\int_{-\infty}^{\infty} \phi(x)p(x)\int_{\mathbb{Z}} \frac{1}{\sqrt{2\pi}}e^{-\frac{1}{2}\left(z^2-2\sqrt{t\kappa_t}\sigma x z+t\kappa_t\sigma^2x^2\right)}\mathrm{d}z\mathrm{d}x \\
&\quad -e^{-rt}K\int_{\mathbb{X}} p(x)\int_{\mathbb{Z}} \frac{1}{\sqrt{2\pi}}e^{-\frac{1}{2}\left(z^2-2\sqrt{t\kappa_t}\sigma x z+t\kappa_t\sigma^2x^2\right)}\mathrm{d}z\mathrm{d}x \\
&= e^{-rT}\int_{\mathbb{X}} \phi(x)p(x)\int_{\mathbb{Z}} \frac{1}{\sqrt{2\pi}}e^{-\frac{1}{2}\left(z-\sqrt{t\kappa_t}\sigma x\right)^2}\mathrm{d}z\mathrm{d}x \\
&\quad -e^{-rt}K\int_{\mathbb{X}} p(x)\int_{\mathbb{Z}} \frac{1}{\sqrt{2\pi}}e^{-\frac{1}{2}\left(z-\sqrt{t\kappa_t}\sigma x\right)^2}\mathrm{d}z\mathrm{d}x.
\end{aligned} \tag{2.77}$$

As we are interested in the expected value where $\xi_t \geq \xi_t^\star$ (or, equivalently, $z \geq z^\star$), Eq. (2.77), in fact, corresponds to

$$\begin{aligned}
C_0 &= e^{-rT}\int_{\mathbb{X}} \phi(x)p(x)\Theta\left(-z^\star+\sqrt{\kappa_t t}\sigma x\right)\mathrm{d}x \\
&\quad -e^{-rt}K\int_{\mathbb{X}} p(x)\Theta\left(-z^\star+\sqrt{\kappa_t t}\sigma x\right)\mathrm{d}x
\end{aligned} \tag{2.78}$$

and

$$\begin{aligned}
P_0 &= e^{-rt}K\int_{\mathbb{X}} p(x)\Theta\left(z^\star-\sqrt{\kappa_t t}\sigma x\right)\mathrm{d}x \\
&\quad -e^{-rT}\int_{\mathbb{X}} \phi(x)p(x)\Theta\left(z^\star-\sqrt{\kappa_t t}\sigma x\right)\mathrm{d}x
\end{aligned} \tag{2.79}$$

for call and put prices, respectively, where $\Theta(\cdot)$ denotes the standard normal cumulative distribution.

Corollary 2.5 *Assume $\phi(X_T)$ is as given in Eq. (2.46).*[13] *Then, explicitly,*

$$\xi^* = \frac{\left(2\left(\ln\frac{K}{S_0}+\frac{\nu^2}{2}T-rt\right)\right)\left(\kappa_t\sigma^2t+1\right)-\nu^2T}{2\kappa_t\sqrt{T}\sigma\nu}. \tag{2.80}$$

[13]It reads $\phi(X_T) = S_0\exp\left[\left(\mu-\frac{1}{2}\nu^2\right)T+\nu\sqrt{T}X_T\right]$, where $X_T \sim \mathcal{N}(0,1)$.

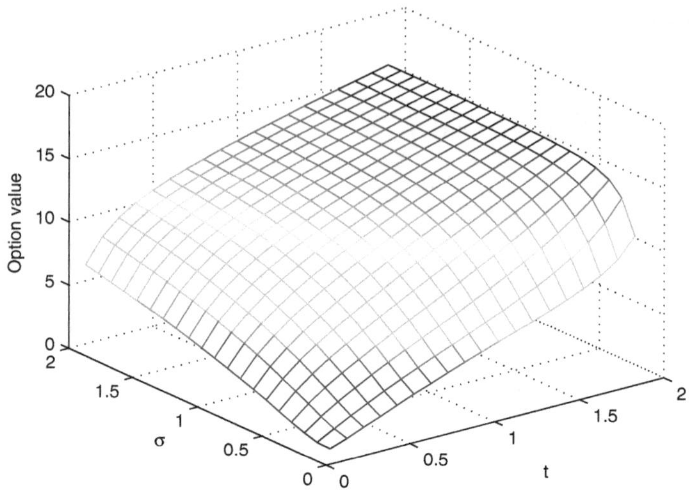

Fig. 2.2 Call option value for ranging pairs (σ, t). Arbitrary parameters: $T = 2$, $\mu = 0.05$, $K = 100$, $\nu = 0.2$, $S_0 = 100$

Figure 2.2 depicts the call option value against a range of values for the information flow rate σ and maturity t, where T is a constant and $\phi(X_T)$ is as in Eq. (2.46). As expected, greater time frame to maturity implies greater chances of price exceeding the threshold K. But, again, how do we interpret faster information flow leading to higher uncertainty and, therefore, higher option prices? At this point, we would like to distinguish between the pattern in which information has been incorporated into prices, and "informativeness" of prices in the sense of [22]. As we shall see in Sect. 2.4 later in this chapter, for any $t \in [0, T]$, when $\sigma_1 \leq \sigma_2$, the relation $h(\phi(X_T)|\xi_t^2) \leq h(\phi(X_T)|\xi_t^1)$ always holds among conditional entropies of $\phi(X_T)$, i.e., ξ_t^2 quickly turns into a less informative signal. Although this might initially seem somewhat contradictory, pricing of contingent claims is more about volatility, i.e., the pattern that information is incorporated, and the variability of signals, i.e., $\mathbb{V}(\xi_t) = (\sigma t)^2 \mathbb{V}(X_T) + t\kappa_t^{-1}$, and, therefore, variability of model prices are increasing in σ, leading to higher option prices.

While the present signal-based framework enables chopping up of the valuation problem into modelling and prediction of market factors, it also reduces the challenge of explaining price variability to determining (e.g., based on data) how fast new information is revealed to the market. Since volatility is latent and cannot be observed directly, it has to be inferred from the data using a certain metric (with squared deviations from the mean being the most common one). Information flow, however, is more intuitive and can be modelled more structurally as well as in a more forward-looking manner.

Figure 2.3 shows the relationship between implied volatility and signal-to-noise ratio of call options written on two selected tickers, namely, `AAPL` and `MSFT`, for different information maturities T. Each curve is an iso-maturity and the vertical

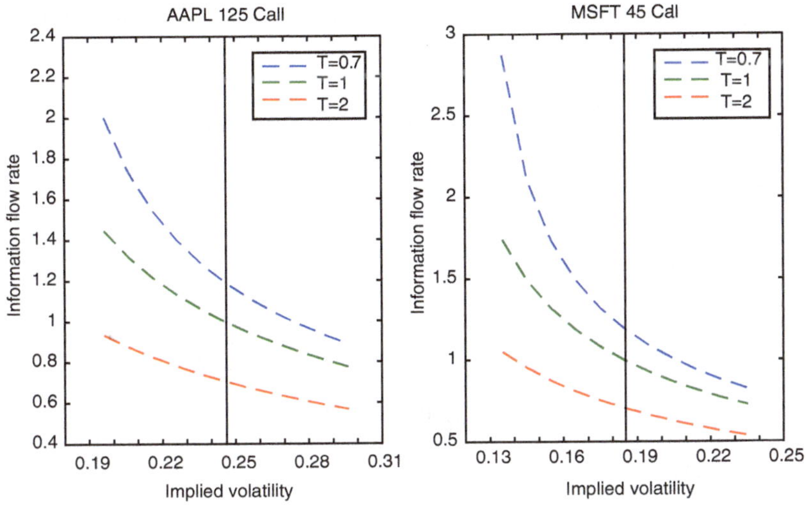

Fig. 2.3 Implied volatility and information flow rate. Options are valued on May 1, 2015, and mature on August 21, 2015

lines show the Black-Scholes implied volatilities on the valuation date. Implied volatility has in fact no financial meaning, other than being an additional degree of freedom to equate the model output to the market reality, and is not sensitive to forward-looking information. We can infer from the figure that information flow rate offers a more intuitive substitute to implied volatility, with substitution rate decreasing as T increases, and allow us to make observations such as "information is flowing more rapidly/slowly to the market," not just "market prices imply a higher/lower volatility."

2.4 An Information-Theoretic Analysis

Another intriguing question at this point would be how much information about $\phi(X_T)$ is carried in signal ξ_t at time t. This would help us measure the change, both w.r.t. time and different values of the flow parameter σ, in the amount of information carried by ξ_t.

We can write the Shannon [31] entropy (which is a special case of Renyí [29] entropy) of ξ_t as

$$h(\xi_t) = -\int_{\Xi} p(\xi_t) \log_2 p(\xi_t) \mathrm{d}\xi_t = -\mathbb{E}_{p(\xi_t)} \left[\log_2 p(\xi_t)\right], \tag{2.81}$$

where $h(\cdot) \geq 0$ almost surely, and $\xi_t \in \Xi$, i.e. the support of ξ_t (see, e.g., [16]). Thus, in general, the greater the variance of ξ_t, the greater its entropy will be.

In the present framework, however, we are more concerned about information in a bilateral sense. In this regard, 'joint' and 'conditional' entropies are defined as

$$h(\xi_t, \phi(X_T)) = -\int_{\Xi}\int_{\mathbb{X}} p(\xi_t, x)\log_2 p(\xi_t, x)\mathrm{d}x\mathrm{d}\xi_t = -\mathbb{E}_{p(\xi_t,x)}\left[\log_2 p(\xi_t, x)\right], \tag{2.82}$$

and

$$\begin{aligned} h(\xi_t|\phi(X_T)) &= -\int_{\mathbb{X}}\left(\int_{\Xi} p(\xi_t|x)\log_2 p(\xi_t|x)\mathrm{d}\xi\right)p(x)\mathrm{d}x \\ &= -\int_{\mathbb{X}}\int_{\Xi} p(\xi_t, x)\log_2 p(\xi_t|x)\mathrm{d}\xi_t\mathrm{d}x \\ &= -\mathbb{E}_{p(\xi_t,x)}\left[\log_2 p(\xi_t|X)\right], \end{aligned} \tag{2.83}$$

respectively, where

$$\begin{aligned} p(\xi, x) &= \frac{\partial^2}{\partial\xi\partial x}\mathbb{Q}\left[(\xi_t < \xi) \cap (\phi(X_T) < x)\right] \\ &= \frac{\partial^2}{\partial\xi\partial x}\mathbb{Q}\left[\xi_t < \xi|\phi(X_T) < x\right]\mathbb{Q}\left[\phi(X_T) < x\right] \\ &= p(\xi|\phi(X_T) = x)p(x) \\ \text{or,} \quad &= p(x|\xi_t = \xi)p(\xi). \end{aligned} \tag{2.84}$$

Using Eq. (2.83), we can work out the following property:

$$\begin{aligned} h(\xi_t|\phi(X_T)) &= -\int_{\mathbb{X}}\left(\int_{\Xi} p(\xi_t|x)\log_2\frac{p(x|\xi_t)p(\xi_t)}{p(x)}\mathrm{d}\xi_t\right)p(x)\mathrm{d}x \\ &= -\int_{\mathbb{X}}\int_{\Xi} p(\xi_t, x)\log_2 p(x|\xi_t)\mathrm{d}\xi_t\mathrm{d}x + \left(-\int_{\Xi} p(\xi_t)\log_2 p(\xi_t)\mathrm{d}\xi_t\right) \\ &\quad -\left(-\int_{\mathbb{X}} p(x)\log_2 p(x)\mathrm{d}x\right) \\ &= h(\phi(X_T)|\xi_t) + h(\xi_t) - h(\phi(X_T)), \end{aligned} \tag{2.85}$$

which directly implies

$$h(\phi(X_T)) - h(\phi(X_T)|\xi_t) = h(\xi_t) - h(\xi_t|\phi(X_T)). \tag{2.86}$$

It is straightforward to see from its definition that, $\forall t \leq T$, ξ_t is of higher entropy (more uncertain) without the knowledge of $\phi(X_T)$ than with it and, therefore, $h(\xi_t) - h(\xi_t|\phi(X_T)) \geq 0$ should hold. This means, in turn, by virtue of the left-hand-side of Eq. (2.86), that $\phi(X_T)$ is of higher entropy (more uncertain) without the knowledge

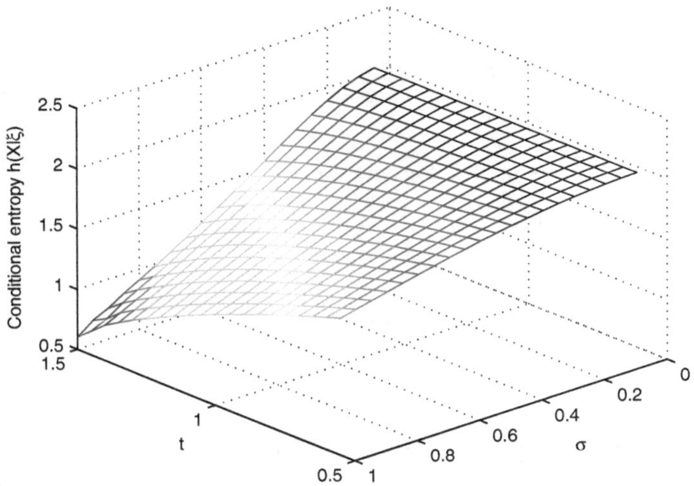

Fig. 2.4 Evolution of conditional entropy $h(\phi(X_T)|\xi_t)$ over time and across information flow rates σ. Arbitrary parameters: $T = 2$, $\phi(X_T) = X_T \sim \mathcal{N}(0, 1)$

of ξ_t than with it otherwise (we've already used this latter property in Sect. 2.3). To illustrate this point further, we depict in Fig. 2.4 the evolution of conditional entropy of $\phi(X_T)$ with respect to ξ_t, i.e., $h(\phi(X_T)|\xi_t)$, for different values of σ. Note that $h(\phi(X_T)|\xi_t)$ decreases both as the signal ξ_t (for a given σ) reveals more information in time, and as its quality, i.e., σ, increases (for a given t).

'Mutual information,' on the other hand, measures the amount of information that ξ_t contains about another random variable $\phi(X_T)$, or vice versa, and corresponds to the reduction in the amount of uncertainty of one variable due to the knowledge of other. Mutual information (I), in fact, corresponds to the 'relative entropy,' or Kullback–Leibler Distance, (D) between $p(\xi, x)$ and $p(\xi)p(x)$, i.e.,

$$\begin{aligned} I(\xi_t, \phi(X_T)) = D(p(\xi, x)||p(\xi)p(x)) &= \int_{\Xi}\int_{\mathbb{X}} p(\xi, x) \log_2 \frac{p(\xi, x)}{p(\xi)p(x)} \mathrm{d}\xi \mathrm{d}x \\ &= \mathbb{E}_{p(\xi,x)}\left[\log_2 \frac{p(\xi, x)}{p(\xi)p(x)}\right], \end{aligned} \tag{2.87}$$

where $I(\xi_t, \phi(X_T)) = D(p(\xi, x)||p(\xi)p(x)) \geq 0$ almost surely.[14] Furthermore, a direct relationship between joint entropy $h(\xi_t, \phi(X_T))$ and mutual information

[14]We complete the definition (2.87) by adding the conditions $0 \ln(0/0) = 0$, $0 \ln(0/p(\cdot)) = 0$, and $p(\cdot) \ln(p(\cdot)/0) = \infty$.

$I(\xi_t, \phi(X_T))$ can be established using Eqs. (2.82) and (2.87):

$$\begin{aligned} I(\xi_t, \phi(X_T)) &= \int_{\Xi}\int_{\mathbb{X}} p(\xi, x)\log_2 p(\xi, x)\mathrm{d}\xi\mathrm{d}x - \int_{\Xi} p(\xi)\log_2 p(\xi)\mathrm{d}\xi \\ &\quad - \int_{\mathbb{X}} p(x)\log_2 p(x)\mathrm{d}x \\ &= h(\xi_t) + h(\phi(X_T)) - h(\xi_t, \phi(X_T)). \end{aligned} \tag{2.88}$$

In the present context, since S_t is an invertible function of ξ_t, we can also show that $I(\xi_t, \phi(X_T)) = I(S_t, \phi(X_T))$ (cf. [12]).

2.5 Single Dividend–Multiple Market Factors

We leave the case where the asset pays multiple cashflows to Chap. 5 and focus on the case where asset pays a single dividend that is, this time, determined by a multiplicity of independent market factors, i.e., $X_T^1, \ldots, X_T^m$, and, therefore, multiple information processes $\xi_t^1, \ldots, \xi_t^m$. Then, S_t is given by

$$\begin{aligned} S_t &= \mathbf{1}_{\{t<T\}} e^{-r(T-t)}\mathbb{E}\left[\phi_T(X_T^1, \ldots, X_T^m)|\xi_t^1, \ldots, \xi_t^m\right] \\ &= \mathbf{1}_{\{t<T\}} e^{-r(T-t)}\int_{\mathbb{X}}\cdots\int_{\mathbb{X}} \phi_T(x_1, \ldots, x_m)\pi_t^1(x_1)\ldots\pi_t^m(x_m)\mathrm{d}x_1\ldots\mathrm{d}x_m, \end{aligned}$$

where, again, $\pi_t^i(x)$'s are the posterior density functions as in Eq. (2.18).[15] Similarly, multi-factor analogue of $\mathrm{d}\phi(X_T)$ in (2.49)[16] is given by

$$\begin{aligned} \mathrm{d}\phi(X_T^1, \ldots, X_T^m) &= \sum_{i=1}^{m} \sigma^i\kappa_t\mathbb{C}\mathrm{ov}_t^i\left(\phi(X_T^1, \ldots, X_T^m), X_T^i\right) \\ &\quad \cdot\left(\kappa_t T^{-1}\left(\xi_t^i - \sigma^i T\mathbb{E}_t\left[\phi(X_T^1, \ldots, X_T^m)\right]\right)\mathrm{d}t + \mathrm{d}\xi_t^i\right), \end{aligned} \tag{2.89}$$

where $\mathbb{C}\mathrm{ov}_t^i := \mathrm{Cov}\left[\phi(X_T^1, \ldots, X_T^m), X_T^i|\xi_t^1, \ldots, \xi_t^m\right]$. We, again, define

$$\mathrm{d}W_t^i \triangleq \kappa_t\left(\xi_t^i T^{-1} - \sigma^i\mathbb{E}_t\left[\phi(X_T^1, \ldots, X_T^m)\right]\right)\mathrm{d}t + \mathrm{d}\xi_t^i \tag{2.90}$$

[15]It reads $\pi_t(x) = p(x)e^{\kappa_t\left(\sigma x\xi_t^{\alpha} - \frac{1}{2}\sigma^2x^2t\right)} / \int_{\mathbb{X}} p(x)e^{\kappa_t\left(\sigma x\xi_t - \frac{1}{2}\sigma^2x^2t\right)}\mathrm{d}x$.

[16]It reads $\mathrm{d}\phi(X_T) = \sigma\kappa_t\mathbb{C}\mathrm{ov}_t\left(\phi(X_T), X_T\right)\left[\kappa_t\left(T^{-1}\xi_t - \sigma\mathbb{E}_t\left[\phi(X_T)\right]\right)\mathrm{d}t + \mathrm{d}\xi_t\right]$.

or, equivalently,

$$W_t \triangleq \xi_t + \int_0^t \kappa_s \left(\xi_s/T - \sigma \mathbb{E}_s \phi(X_T^1, \ldots, X_T^m)\right) \mathrm{d}s. \tag{2.91}$$

Equation (2.90) enables us to simplify Eq. (2.89) as

$$\mathrm{d}\phi(X_T^1, \ldots, X_T^m) = \sum_{i=1}^m \sigma^i \kappa_t \mathbb{C}\mathrm{ov}_t^i \mathrm{d}W_t^i. \tag{2.92}$$

Since $S_t = \mathbf{1}_{\{t<T\}} e^{-r(T-t)} \phi(X_T^1, \ldots, X_T^m)$, in a similar sense to Eq. (2.28), we can write the dynamics $\mathrm{d}S_t$ as

$$\mathrm{d}S_t = \mu S_t \mathrm{d}t + \sum_{i=1}^m \Lambda_t^i \mathrm{d}W_t^i, \tag{2.93}$$

where we again used the definition $\Lambda_t^i := e^{-r(T-t)} \sigma^i \kappa_t \mathbb{C}\mathrm{ov}_t^i$. This implies that the overall absolute volatility at time t is

$$\Lambda_t = \left(\sum_{i=1}^m (\Lambda_t^i)^2 \right)^{1/2}. \tag{2.94}$$

Equation (2.94) is quite telling in the sense that it enables us to decompose the absolute volatility at time t into its unhedgeable stochastic volatility components (see [24]).

References

1. Back K, Pedersen H (1998) Long-lived information and intra-day patterns. J Financ Mark 1:385–402
2. Bedini M, Buckdahn R, Engelbert H-J (2016) Brownian bridges on random intervals. arXiv:1601.01811v1
3. Belen S, Kropat E, Weber G-W (2009) On the classical Maki-Thompson rumour model in continuous time. CEJOR 19(1):1–17
4. Bhar R (2010) Stochastic filtering with applications in finance. World Scientific, Singapore
5. Bommel J (2003) Rumors. J Financ LVIII(4):1499–1519
6. Brody D, Hughston L (2013) Lévy information and the aggregation of risk aversion. Proc R Soc Lond A Math Phys Sci 469(2154). http://rspa.royalsocietypublishing.org/content/469/2154/20130024
7. Brody D, Law Y (2015) Pricing of defaultable bonds with random information flow. Appl Math Financ 22(5):399–420
8. Brody D, Hughston LP, Macrina A (2007) Beyond hazard rates: a new framework for credit-risk modelling. In: Advances in mathematical finance. Applied and numerical harmonic analysis, chapter III. Birkhäuser, Boston, pp 231–257

9. Brody D, Hughston LP, Macrina A (2008) Dam rain and cumulative gain. Proc Math Phys Eng Sci 464(2095):1801–1822
10. Brody D, Hughston L, Macrina A (2008) Information-based asset pricing. Int J Theor Appl Financ 11(1):107–142
11. Brody D, Davis M, Friedman R, Hughston L (2009) Informed traders. Proc R Soc A 465:1103–1122
12. Brody D, Hughston L, Macrina A (2011) Modelling information flows in financial markets. Advanced mathematical methods for finance. Springer, Berlin, pp 133–153
13. Brody D, Hughston L, Yang X (2013) Signal processing with Lévy information. Proc R Soc Lond A469:20120433
14. Çaglar M, Sezer AD (2014) Analysis of push-type epidemic data dissemination in fully connected networks. Perform Eval 77:21–36
15. ChunXia Y, Sen H, BingYing X (2012) The endogenous dynamics of financial markets: interaction and information dissemination. Physica A 391(12):3513–3525
16. Cover TM, Thomas JA (2006) Elements of information theory, 2nd edn. Wiley, Hoboken
17. Girsanov I (1960) On transforming a certain class of stochastic processes by absolutely continuous substitution of measures. Theory Probab Appl 5(3):285–301
18. Hoyle A (2010) Information-based models for finance and insurance, Ph.D. Thesis, Department of Mathematics, Imperial College London, London SW7 2AZ, United Kingdom
19. Hoyle A, Hughston L, Macrina A (2011) Lévy random bridges and the modelling of financial information. Stoch Process Appl 121:856–884
20. Hughston L, Macrina A (2008) Information, inflation, and interest. In: Advances in mathematics of finance, vol 83. Banach Center Publications, Warszawa
21. Kalman R (1960) A new approach to linear filtering and prediction problems. J Basic Eng (Ser D) 82:35–45
22. Kyle A (1985) Continuous auctions and insider trading. Econometrica 53(6):1315–1335
23. Liang X (2013) The Liang-Kleeman information flow: theory and applications. J Entropy 15(1):327–360
24. Macrina A (2006) An information-based framework for asset pricing: X-factor theory and its applications, Ph.D. Thesis, University of London
25. Macrina A, Parbhoo P (2010) Security pricing with information-sensitive discounting, Cornell University Library ArXiv e-prints: 1001.3570
26. Mengütürk L (2012) Information-based jumps, asymmetry and dependence in financial modelling, Ph.D. Thesis, Department of Mathematics, Imperial College London, London SW7 2AZ, United Kingdom
27. Moll VH (2014–2015) Special integrals of Gradshteyn and Ryzhik: the proofs. Monographs and research notes in mathematics, vol 1–2. Chapman and Hall/CRC Press, London/Boca Raton
28. Øksendal B (1998) Stochastic differential equations: an introduction with applications, 5th edn. Springer, Berlin
29. Renyi A (1984) A diary on information theory. Wiley series in probability and mathematical statistics. Wiley, New York
30. Rutkowski M, Yu N (2007) An extension of the Brody–Hughston–Macrina approach to modeling of defaultable bonds. Int J Theor Appl Financ 10(3):557–589
31. Shannon C (1948) A mathematical theory of communication. Bell Syst Techn J 27:379–423
32. Stein C (1981) Estimation of the mean of a multivariate normal distribution. Ann Stat 9(6):1135–1151
33. Yang X (2013) Information-based commodity pricing and theory of signal processing with Lévy information, Ph.D. Thesis, Department of Mathematics, Imperial College London and Shell International, London SW7 2AZ, United Kingdom

Chapter 3
A Signal-Based Heterogeneous Agent Network

Trade can occur on purely informational causes. In [6], for example, we are shown that there are situations in which both parties are strictly better off under a trade executed solely on the basis of their individual information.[1] The literature on the dynamics of heterogenous markets is still in its infancy but actively developing. Indeed, one can be overwhelmed by the task of handling a very broad spectrum of aspects where agent-level heterogeneity can arise, such as risk aversion levels, degrees of rationality, patience, beliefs, and information gathering, processing skills, and so on.

A detailed classification of different market microstructure models, on the other hand, is given in [10] which is beyond the scope of the present chapter. However, we start with a review of the selected literature. In this regard, Table 3.1 provides a summary overview of the literature on equilibrium information-based agent networks.

Perhaps one of the earliest sequential (discrete) trade models is the one described in the work of Glosten and Milgrom (cf. [15]), where an attempt is made to explain bid-ask spread as a purely informational phenomenon that is believed to be arising from adverse selection behaviour encountered by less-informed traders. The informational properties of transaction prices and the reaction of the spread to market-generated as well as other public information is also investigated. One of the interesting implications of this model is the possibility of market shutdowns due to severe informational inefficiencies. This is similar to the "lemons problem" of Akerlof [2]. The informational content of prices and the value of extra information to the holder are also examined in the work of Kyle (cf. [18]) through sequential as well as continuous auction models. Moreover, the latter two seem to converge as the trading interval gets smaller. One interesting result of the model discussed in [18], and to a certain extent in [15], is that modelling innovations as functions of quantities traded is found to be consistent with modelling price innovations as the

[1]Contrary to, e.g., [16, 19].

N.S. Aydın, *Financial Modelling with Forward-looking Information*, Contributions to Management Science, DOI 10.1007/978-3-319-57147-8_3

Table 3.1 Selected literature on the analysis of heterogeneous information- or belief-based market dynamics and equilibria

Author	Model	Agents	Signal structure	Signal deadline	Payoff distribution	Implications
Glosten and Milgrom [15]	Sequential, market order, unit quantity, individual order	Informed (m), liquidity (m), competitive specialist (m)	Immediate access, long-lived	Fixed to trading period	Continuous (normal)	Differences on quantity and quality of possessed information help explain existence and magnitude of (protracted) spreads
Kyle [18]	Sequential & continuous, market order, optimal quantity	Informed (s), noise (m), competitive market maker (m)	Immediate access, long-lived	Fixed beyond trading period	Continuous (normal)	Optimal strategy exists to incorporate extra information smoothly into prices without affecting price volatility, and maximise profits
Admeti and Pfleider [1]	Sequential Kyle-type auctions, market orders	Informed (m), liquidity (m), market-maker (s), all risk-neutral	Short-lived, explicit	One-period	Continuous (random walk)	Clustering of liquidity trading induces clustering of information driven price changes
Back [3]	Continuous, optimal quantity	Informed (s), uninformed (m), comp. market maker (m)	Immediate access, long-lived	Fixed to trading period	Continuous (general)	Kyle model equilibrium exists for general continuous distributions without recourse to filtering theory
Back and Pedersen [5]	Continuous, optimal quantity	Informed (s), uninformed (m), risk-neutral competitive market maker (m)	Long-lived, explicit	Fixed to trading period	Continuous (general)	Information flow pattern is irrelevant given the total amount of information. Intensity of information use follows intensity of liquidity trading
Back and Baruch [4]	Continuous, market order,	Informed (s), uninformed (m), competitive market maker (m)	Immediate access, filtration	Random, equal to trading period	Discrete	Continuous version of Glosten-Milgrom converges to continuous Kyle model

Caldentey and Stachetti [12]	Sequential & continuous, market order	Informed (s), uninformed (m), market maker (s)	Long-lived, not explicit (may not reveal X fully)	Random trading horizon	Continuous (random walk)	All extra information is used before endogenous T and then, until random deadline, it is passed on to market immediately. Larger the variance of signal, smaller the T, larger the rents
Bond and Eraslan [6]	One-period, endogenous dividend	Risk-neutral agents (p)	Not explicit, $\in \mathbb{R}$	One-period	Binary	Trade based purely on informational differences is possible. Common knowledge of strict gains from information-based trade exists
Brown and Rogers [9]	Sequential & continuous, belief-based	Central-planner (s), agents (m)	Not explicit, belief process Λ_t	Fixed to trading period	Endowment process	Observational equivalence (i.e., equilibria & portfolio choices) of private information and diverse beliefs models is established

Notes: (s): single, (p): pair, (m): multiple (> 2)

consequence of new information arrivals. The ‘informativeness of prices’ (which is complementary to the amount of information which is yet to be incorporated into prices) in the context of [18] refers to the error variance of future dividend given the market clearing price. The question is how intensively the agent, given his superior signal, should trade over time to maximise his profit given his actions might disturb the market (i.e., prices and depth). This model is later on extended in [3] to general continuous distributions for the dividend. Then, a modified version of [15] in continuous time, where ‘bluffing’ (i.e., mixed strategy) is also allowed, is shown to converge to, again, a modified version of [18] with a random signal deadline in [4]. A rather game-theoretic approach to signal-based trade is taken in [6] where, this time, the dividend is let endogenously be determined by the action of the agent and its correspondence with the realised fundamental. The signals, in this case, are related to the action that needs to be taken. A sufficient level of signal precision is found to be necessary and sufficient for establishing the case where both seller and buyer are better off from trade in expectation (referred to as “common knowledge of gains from trade” in [6]), which is the equilibrium.

So far, there is no explicit mention of the dynamics of information flow, which is the subject of heterogeneity, and it is understood to be an ‘immediate access’ to a publicly unknown value $\phi(X_T)$ without any noise component. Building on [3, 18], a learning component is added in [5]. This means the signals are now long-lived with a signal-to-noise varying in time. Although this made possible the mentioning of information ‘flow’ in its true meaning, the interpretation of ‘learning’ through signal in [5] is slightly different in that when the noise-to-signal, i.e., reciprocal of signal-to-noise, is large, this means the agent is learning a lot. Yet, interestingly, given the total amount of information disparity in favour of the more informed, the pattern in which the information flows is found to be rather irrelevant in equilibrium. Later on, the long-lived signal process is associated with a exponential distributed random deadline (as in [4] earlier) in [12]. In fact, a random deadline changes the way the strategies for exploiting extra information are structured in various ways, with one way being that agents do not rush to unload their information before it becomes useless and, accordingly, trade frantically as deadline approaches. Backward induction methods of dynamics programming are also rendered inapplicable.

Perhaps the most interesting alternative to the models of ‘diverse information’ models (where agents do generally share the same probability measure but work over distinct probability spaces) are those of ‘diverse beliefs’. One way to account for the diversity of beliefs is through equivalent (i.e., defined over the same filtered probability space) probability measures which reflect agents’ personal beliefs on the true value of the dividend, as in [9]. This is maintained by likelihood ratio martingales (or, density processes). Interestingly, the equivalence of the latter two models is established, even without a particular choice of explicit signal structure for private information. And, not so interestingly, the greater the diversity of beliefs, the larger the volume of trade is. A similar approach is found in [13] where an equilibrium is established in terms of ‘surviving agents.’ In a belief-heterogeneous market, the surviving agent is found to be the one who is the most rational. Last but not least, in cognisance of the important role played by dynamic optimisation in approaching heterogeneous financial market equilibrium problems, we underline two recent accounts of the latter, i.e. [11, 14], where how, in a market of two agents with heterogeneous characteristics, equilibria for various quantities can be found by means of a single backward-induction algorithm is vividly shown.

The approach in the rest of this chapter to being informationally (dis)advantageous is analogous to the one in [7, 8]: we do not view the state of being informationally (dis)advantageous as (not) having immediate access to the future value of a variable which is unknown to the public information. We rather view it as having access to efficient streams of information or, equivalently (cf. [1]), being more capable to compile and process large and complex datasets out of publicly available information. Both of these are associated with a higher σ of ξ in the context of Chap. 2. Yet, in the sense of [9], the present framework can also be seen as a diverse belief model where beliefs are shaped in time by information itself.

3.1 Model Setup

We assume that there is a pure dealership market comprising risk-neutral agents with heterogeneous informational access. For simplicity, and w.l.o.g., we assume there is a pair of agents, $j = 1, 2$, with access to the filtrations ξ_t^1, ξ_t^2, and a single risky asset with payoff $\phi(X_T)$ not being measurable w.r.t. $\mathcal{F}_t^{\xi}$, $t < T$. We also assume $\sigma_1 < \sigma_2$, i.e., agent 2 is informationally more susceptible than agent 1. In our dynamical model, for simplicity of analysis, we suppose that agents trade with each other futures contracts on the single risky asset at sequential auction times $t_i \in [0, T]$ for $i = 1, 2, \ldots, m$, without any intertemporal consumption and exogenous wealth. Both agents simply follow a buy-and-hold strategy. In this setup, execution of trades, besides a potential profit or loss, results in two things. First, they help, e.g., the central-planner, consolidate[2] information sets of agents at time t_i to have a joint information bundle $\bar{\sigma}^c = \sigma(\xi^1, \xi^2)$. Second, the competitive market price will be discovered immediately. Below we will analyse the latter two separately. Limit orders are cleared by a Walrasian matching engine (as in [11]), which can be deemed a central-planner in the context of [9] or a group of competitive market makers. The central-planner aims solely to maximise the overall expected profit (or, utility) of agents.

We also note that, for any given t and a priori density $p(x)$, the price is a function of ξ and σ, i.e., $S_t = S(\xi, \sigma)$. This means, if S_t is observed, then one needs to know σ to be able to back out ξ_t. Without knowledge σ, the observer cannot infer how reliable an observed sample of ξ_t is.

[2]Here we emphasise the term 'consolidate,' since how information is consolidated will be one of the key questions in our algorithm.

Moments before the sequential auction time t_i, agents, having observed their signals, submit to the central planner the bid and ask prices at which they are willing to trade. One key property of our model is that an agent may not necessarily know his signal is superior (i.e., agnostic) and the agents will be able to infer each other's prices, and also information (unless they are 'omitters', as described below), when, and only if, a price match occurs and a clearing price is set. Otherwise, limit orders are kept with the auction engine (i.e., closed limit order trading book). This also rules out 'bluffing' (cf. [4, 12]).

Individual bid and ask prices are based on the signal-implied prices worked out by virtue of Eq. (2.19) and trade occurs whenever

$$\varsigma^- S_t^1 \geq \varsigma^+ S_t^2 \quad \text{or} \quad \varsigma^- S_t^2 \geq \varsigma^+ S_t^1, \tag{3.1}$$

with ς^- and ς^+ being the constant bid and ask multipliers, respectively, where $\varsigma^- \leq 1$ and $\varsigma^+ \geq 1$. Obviously, if Eq. (3.1) holds with equality, i.e., if $\varsigma^- S_t^1 = \varsigma^+ S_t^2$ or $\varsigma^- S_t^2 = \varsigma^+ S_t^1$, then the market price S_t^* will be discovered directly. In case of an inequality, under risk-neutrality assumption, the market will clear at the mid-price[3]

$$S_t^* = \frac{\varsigma^- S_t^1 + \varsigma^+ S_t^2}{2} \quad \text{or} \quad \frac{\varsigma^- S_t^2 + \varsigma^+ S_t^1}{2}. \tag{3.2}$$

The initial contract holdings of agents, as denoted by $\theta_0^j, j \in \{1, 2\}$, are set to 0. Here θ_t^j denotes the total time-t net contract stock held by agent j. We also define $\theta_t := \sum_j \theta_t^j$ as the total net contract 'stock' held by the central clearing at time t. Accordingly, total net order 'flow' at time t should be $\Delta\theta_t$ which is given by

$$\Delta\theta_t = \sum_j \Delta\theta_t^j = \sum_j q_t^j = q_t \tag{3.3}$$

for some trading process $(q_t^j)_{0 \leq t \leq T}$, given by

$$q_t^j := \begin{cases} q_t^{j+}, & S_t^j > S_t^*, \\ q_t^{j-}, & S_t^j < S_t^*, \\ 0, & \text{otherwise}, \end{cases} \tag{3.4}$$

with $q_t^{j+} > 0$, $q_t^{j-} < 0$. Market clearing conditions imply $q_t = \sum_j q_t^j = 0$ and, therefore, $\theta_t = 0$, $\forall t \in [0, T]$. Now we define the increasing process $(s_t)_{t \in [0,T]}$, i.e., the time of the last trade prior to time t, as follows:

$$s_t = \sup\{s : s < t, |q_s^j| > 0\}. \tag{3.5}$$

[3]In [11, p. 7], the authors elegantly elaborate why the real-world interpretation of the price posted by a Walrasian auctioneering computer is the bid-ask midpoint.

It is apparent that s_t is 0 if $t = 0$, or if $t > 0$ and $q_s^j = 0 \; \forall s \in [0, t)$. The ex-post (i.e., at contract expiry) profit/loss of agent j coming from time t transaction can be written as

$$\Pi_t^j = \mathbf{1}_{H \cap \{S_t^j > S_t^*\}} q_t^{j+} (X - S_t^*) + \mathbf{1}_{H \cap \{S_t^j < S_t^*\}} q_t^{j-} (X - S_t^*) + \tag{3.6}$$

$$\mathbf{1}_{L \cap \{S_t^j < S_t^*\}} q_t^{j-} (X - S_t^*) + \mathbf{1}_{L \cap \{S_t^j > S_t^*\}} q_t^{j+} (X - S_t^*)$$

$$\text{(or, simply)} = q_t^j (X - S_t^*), \tag{3.7}$$

where S_t^* is as in Eq. (3.2), and H and L denote high- and low-type markets, respectively (cf. [18]). Equation (3.6) is based on the correspondence of signal and reality. Market clearing conditions again will require $\Pi_t = \sum_j \Pi_t^j = 0 \; \forall t \in [0, T]$.

3.2 Numerical Analysis

We now present some numerical results based on the setup above. Let $|q_t^j| \in \{0, 1\}$ and assume, in this first scenario, that both agents are "omitters" (or, "stubborn bigots" of [9]) who never change their mind and simply execute trades according to the following recurring procedure: (1) Observe signal ξ_t^j. (2) Quote signal-based bid and ask prices. (3) Let the central-planner determine—using the pre-announced and legally binding matching rule (3.2)—the trade direction, if any, and the transaction price (which are then revealed to the agents). Note that agents execute trades "without" learning from each other—who could, otherwise, update their likelihoods $p(\xi|x)$ as we will see later on—and continue to rely solely on their own information sources.

In Fig. 3.1, where the true fundamental value of X is set to 1, we illustrate one possible path of such a scenario. Despite a bid-ask margin, occurrence of trade is highly likely in this case as agents do not learn from each other and as personal value judgements diverge. The informationally more (less) susceptible agent, though unknowingly, keeps trading in the right (wrong) direction due to superiority (inferiority) of his signal. Note from Fig. 3.1 that even after the agent with better signal discovers the asset's true value (around auction 5), he is still able to execute profitable trades thanks to the matching rule. Figure 3.2, on the other hand, shows the profit-and-loss (P&L) results of such a scenario for each time step averaged over 10^3 simulations, where the number of auctions is increased to 100. We note at the first glance that the qualitative behaviour of the P&L agrees with the qualitative behaviour of the magnitude of extra information held, presented through our information-theoretical analysis in Sect. 2.4 of Chap. 2 of this book, as well as that in [7].

On an additional note, when multiple (>2) agents with various informational capabilities are involved in the market, our numerical results presented in Fig. 3.3 suggest that, while P&L continue to agree with the qualitative behaviour of the

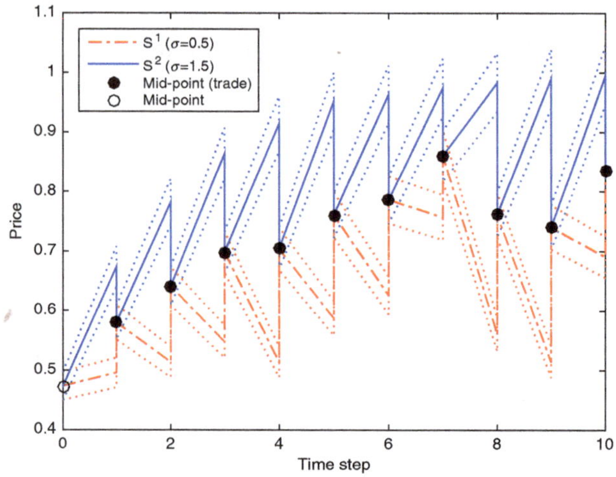

Fig. 3.1 Sample evolution of information-based transaction prices in scenario 1 ($|q_t^j| \in \{0, 1\}$). Arbitrary parameter values: $T = 1$, $\Delta t = 1/10$, $r = 0.05$, $\sigma \in [0.5, 1.5]$, and $\phi(X_T) \in \{0, \bar{1}\}$ (i.e., true value is set to 1) with $p_0(x) \in [0.5, 0.5]$. The *dotted lines* are bid and ask prices based on $S\varsigma^-$ and $S\varsigma^+$, respectively, with $\varsigma \in [0.95, 1.05]$

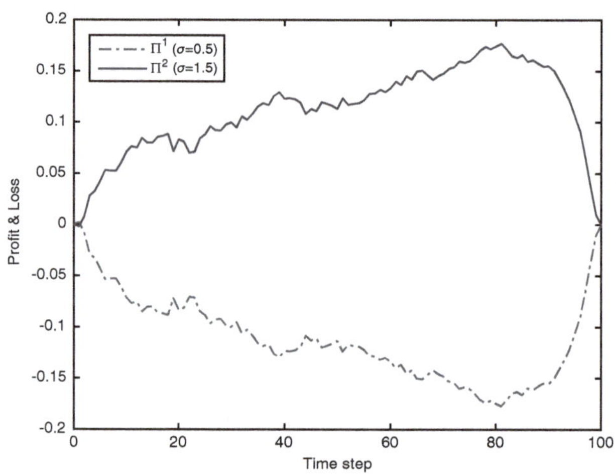

Fig. 3.2 Evolution of information-based transaction P&L averaged over 10^3 path simulations and based on parameters from Fig. 3.1, except that $\Delta t = 1/100$

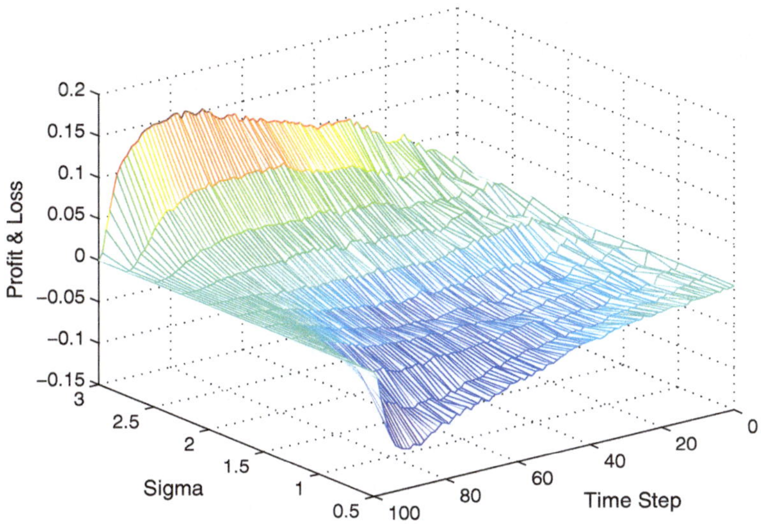

Fig. 3.3 Evolution of information-based transaction P&L of multiple agents averaged over 10^3 path simulations and based on parameters given in Fig. 3.1, except that $\Delta t = 1/100$

magnitude of extra information held by the agent, it is also distributed between agents proportional to the quality of their signal (particularly once the differential informational reaches an adequate level).

Yet, the exchanges generally do not operate quite this way. A more realistic scenario would be that agents are "attentive" and infer their counterpart's posterior $\pi^j_{s_t}(x)$, and, therefore, likelihood $p(\xi^j_{s_t}|x)$, from their price quote at time s_t. This would mean having partial access to a larger σ-algebra, $\bar{\sigma}(\xi_{s_t})$, generated by the join[4] of $\sigma(\xi^1_{s_t})$ and $\sigma(\xi^2_{s_t})$, i.e.,

$$\bar{\sigma}(\xi^j_{s_t}) = \sigma(\xi^1_{s_t}) \vee \sigma(\xi^2_{s_t}). \tag{3.8}$$

Once agent j gains partial access to $\bar{\sigma}(\xi^j_{s_t})$, he updates his posterior from $\pi^j_{s_t}(x)$ to $\bar{\pi}^j_{s_t}(x)$ (by updating $p(\xi^j_{s_t}|x)$ to $\bar{p}(\xi^j_{s_t}|x)$, i.e., the effective likelihood), which will be again of the form

$$\bar{\pi}^j_{s_t}(x) = \frac{p(x)\bar{p}(\xi^j_{s_t}|x)}{\int_{\mathbb{X}} p(x)\bar{p}(\xi^j_{s_t}|x)\mathrm{d}x}. \tag{3.9}$$

[4]Note that the union of collection of sigma algebras is not always a σ-algebra or even an algebra.

Note that we intentionally avoid the notation $p(\xi^1_{s_t}, \xi^2_{s_t}|x)$, and use $\bar{p}(\xi^j_{s_t}|x)$ instead, so as not to mean that one party's signal is directly observable to the other at the last auction time s_t (which is also not needed). By virtue of bi-dimensional normal density, $\bar{p}(\xi^j_{s_t}|x)$ or, effectively, $p(\xi^1_{s_t}, \xi^2_{s_t}|x)$, in Eq. (3.9) can be written in the form

$$\begin{aligned} p(\xi^1_{s_t}, \xi^2_{s_t}|x) = & \frac{1}{2\pi(s_t/\kappa_{s_t})\sqrt{1-\rho^2}} \\ & \cdot \exp\left(-\frac{1}{2}\frac{(\xi^1_{s_t} - \sigma_1 x s_t)^2}{(1-\rho^2)s_t/\kappa_{s_t}}\right) \\ & \cdot \exp\left(-\frac{1}{2}\frac{-2\rho(\xi^1_{s_t} - \sigma_1 x s_t)(\xi^2_t - \sigma_2 x s_t)}{(1-\rho^2)s_t/\kappa_{s_t}}\right) \\ & \cdot \exp\left(-\frac{1}{2}\frac{(\xi^2_{s_t} - \sigma_2 x s_t)^2}{(1-\rho^2)s_t/\kappa_{s_t}}\right), \end{aligned} \tag{3.10}$$

with $|\rho| < 1$ denoting the correlation between $\xi^1_{s_t}$ and $\xi^2_{s_t}$ conditional on x.[5]

Finally, we note that, in the present setup, the effective information $\bar{\sigma}(\xi^j_{s_t})$ can be worked out only after (not before) the trade at time s_t, which renders it literally 'useless' until the present auction at time t. Therefore, before submitting an order at time t, having observed a new signal ξ^j_t, the agent will need to update his effective information to $\bar{\sigma}(\xi^j_t) = \sigma(\xi^1_t) \vee \sigma(\xi^2_{s_t})$ (e.g., for agent 1) or $\sigma(\xi^1_{s_t}) \vee \sigma(\xi^2_t)$ (in the case of agent 2). Also, since ξ_t is Markovian, for an agent, partially accessing the signal sample $\xi^j_{s_t}$ of his counterpart will be as valuable as partially accessing his entire signal history $(\xi^j_s)_{s \le s_t}$. Accordingly, right before the auction at time t, the 'useful' effective likelihood $\bar{p}$ for agent 1 will be

$$\begin{aligned} p(\xi^1_t, \xi^2_{s_t}|x) = & \frac{1}{2\pi\sqrt{t/\kappa_t}\sqrt{s_t/\kappa_{s_t}}\sqrt{1-\hat{\rho}^2}} \\ & \cdot \exp\left(-\frac{1}{2}\frac{(s_t/\kappa_{s_t})(\xi^1_t - \sigma_1 x t)^2}{(1-\hat{\rho}^2)(t/\kappa_t)(s_t/\kappa_{s_t})}\right) \\ & \cdot \exp\left(-\frac{1}{2}\frac{-2\hat{\rho}(\xi^1_t - \sigma_1 x t)(\xi^2_{s_t} - \sigma_2 x s_t)\sqrt{(t/\kappa_t)}\sqrt{(s_t/\kappa_{s_t})}}{(1-\hat{\rho}^2)(t/\kappa_t)(s_t/\kappa_{s_t})}\right) \\ & \cdot \exp\left(-\frac{1}{2}\frac{(t/\kappa_t)(\xi^2_{s_t} - \sigma_2 x s_t)^2}{(1-\hat{\rho}^2)(t/\kappa_t)(s_t/\kappa_{s_t})}\right), \end{aligned} \tag{3.11}$$

[5]Note that when $\rho = 1$, assuming $\sigma_1 \neq \sigma_2$, the central planner will only need to solve two linear equations, with X and β_{t_i-} being the two unknowns, to get instant access to X.

where we used the relation $\beta^1_{s_t} = \rho\beta^2_{s_t} + \sqrt{1-\rho^2}\bar{\beta}_{s_t}$, with $\beta^2_{s_t} \perp\!\!\!\perp \bar{\beta}_{s_t}$, to find $\hat{\rho}$, i.e., the new correlation structure between ξ^1_t and $\xi^2_{s_t}$ given x, as follows:

$$\begin{aligned}\hat{\rho} &= \frac{\mathbb{C}\mathrm{ov}(\beta^1_t, \beta^2_{s_t})}{\sigma_{\beta^1_t}\sigma_{\beta^2_{s_t}}} \\ &= \frac{\mathbb{C}\mathrm{ov}(\rho\beta^2_t + \sqrt{1-\rho^2}\bar{\beta}_t, \beta^2_{s_t})}{\sqrt{t/\kappa_t}\sqrt{s_t/\kappa_{s_t}}} \\ &= \rho\frac{s_t/\kappa_t}{\sqrt{t/\kappa_t}\sqrt{s_t/\kappa_{s_t}}} = \rho\sqrt{\frac{s_t}{t}\frac{\kappa_{s_t}}{\kappa_t}},\end{aligned} \tag{3.12}$$

with ρ being same as in Eq. (3.10). We note that $\hat{\rho}$ is a decreasing function of time, as expected, and also that, when $\hat{\rho} = 0$, Eq. (3.11) simply reduces to

$$\begin{aligned}p(\xi^1_t, \xi^2_{s_t}|x) &= \frac{1}{2\pi\sqrt{t/\kappa_t}\sqrt{s/\kappa_{s_t}}} \\ &\cdot \exp\left(-\frac{1}{2}\frac{(s_t/\kappa_s)(\xi^1_t - \sigma_1 xt)^2}{(t/\kappa_t)(s_t/\kappa_{s_t})}\right) \\ &\cdot \exp\left(-\frac{1}{2}\frac{(t/\kappa_t)(\xi^2_{s_t} - \sigma_2 xs_t)^2}{(t/\kappa_t)(s_t/\kappa_{s_t})}\right),\end{aligned} \tag{3.13}$$

which also reduces to $p(\xi^1_t|x)$ when $s_t = 0$ (no trade). The signal-based price of agent j, S^j_t, is then given by

$$S^j_t = \mathbb{E}\left[X|\bar{\sigma}(\xi^j_t)\right]. \tag{3.14}$$

Accordingly, the new trading procedure is as follows: (1) Observe signal ξ^j_t. (1a) Work out $\bar{\sigma}(\xi^j_t)$. (2) Quote signal-based bid and ask based on effective information. (3) Let the central-planner do his work (same as (3) above).

One realisation of this second scenario is depicted in Fig. 3.4. At the first glance, learning seems to have decreased profit margins substantially (i.e., to a level where they are often eaten up by the spread, preventing trade). In Fig. 3.5, we again show average stepwise P&L of agents over 10^3 realisations. It is apparent from the figure that the informationally more susceptible agent is no more able to extract rents that are as large as in the first scenario (see Fig. 3.2), although he is still able to maintain some modest profits. His ability to maintain modest profits is most likely due to the lag in the learning process as there is still a room for the superior signal to provide the agent receiving it with extra information in-between auctions. The huge difference between the outcomes of two scenarios, i.e., "omitter" and "attentive", implies that, when each agent deems his own signal superior, there might exist

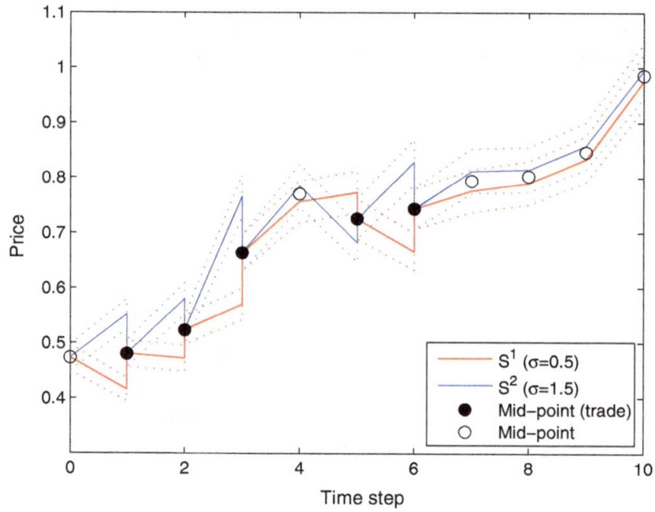

Fig. 3.4 Sample evolution of information-based transaction prices along a sample path in scenario 2 ($|q_t^j| \in \{0, 1\}$). Arbitrary parameter values: $T = 1$, $\Delta t = 1/10$, $r = 0.05$, $\sigma \in [0.5, 1.5]$, and $\phi(X_T) \in \{0, 1\}$ with $p_0(x) \in [0.5, 0.5]$. The *dotted lines* are bid and ask prices based on $S\varsigma^-$ and $S\varsigma^+$, respectively, with $\varsigma \in [0.95, 1.05]$

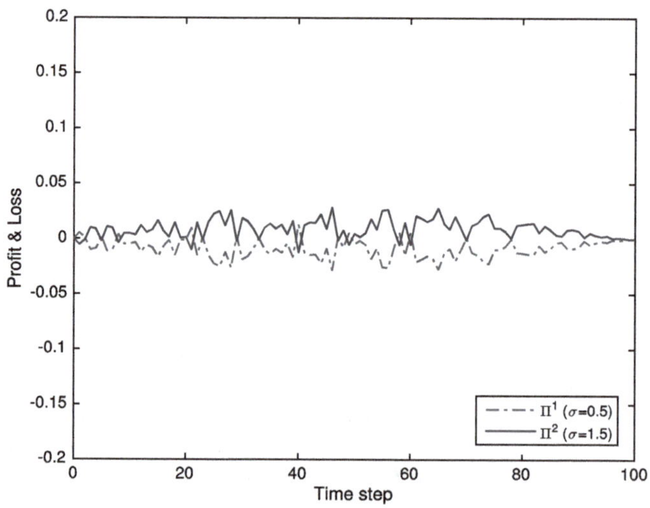

Fig. 3.5 Evolution of information-based transaction P&L averaged over a series of 10^3 path simulations and based on parameters given in Fig. 3.4, except that $\Delta t = 1/100$

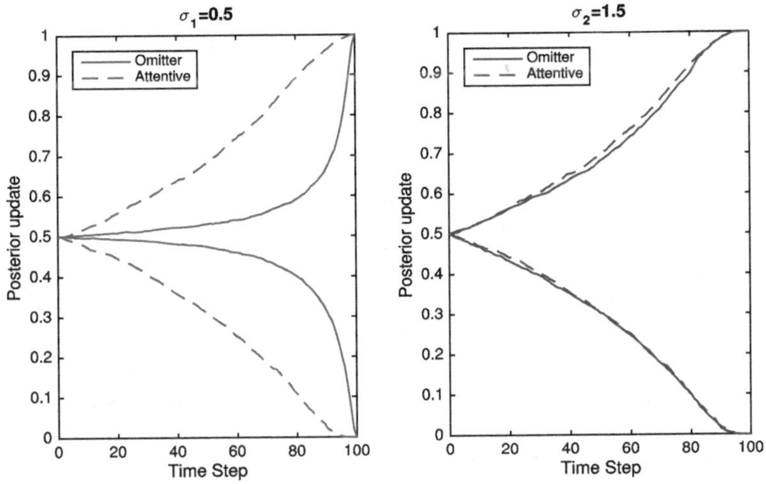

Fig. 3.6 Learning process: Bayesian updating of posteriors $\pi_t^j(x)$ averaged over 10^3 path simulations and based on parameters given in Fig. 3.4, except that $\Delta t = 1/100$

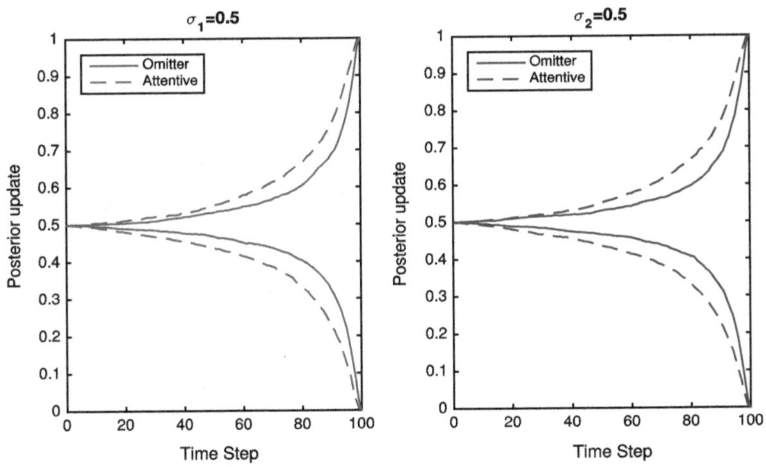

Fig. 3.7 Learning process: Bayesian updating of posteriors $\pi_t^j(x)$ averaged over 10^3 path simulations and based on parameters given in Fig. 3.4, except that $\Delta t = 1/100$

optimal strategies where agents can still be "attentive" but, this time, choose which time to reveal their information through trade.

To conclude this section, we compare, in Figs. 3.6 and 3.7, the impact of allowing mutual learning on the speeds at which the two agents discover the true fundamental value of the asset. In the case where the differential between information flow speeds is high (refer to Fig. 3.6), learning seems to work more in favour of the agent with less superior signal with little or no benefit to the agent with a superior signal, whereas, when the differential is minimal (cf. Fig. 3.7), both agents equally benefit from sharing their information via trading.

3.3 Signal-Based Optimal Strategy

The P&L figures provided in Sect. 3.1 are ex-post, i.e., calculated at the terminal date. In reality, when they trade, agents do so based on their signal-based expectations about the true fundamental value to be revealed at time T. They learn whether their earlier trades in futures contracts turned out to be a profit or loss again at time T. This, in fact, establishes the main argument which calls for the existence of optimal choices of trading times which maximise their signal-based expected profits: both agents believe that their trades will make them better off (or, there exists 'a common knowledge of gains from trade' in the sense of [6]). Throughout this section, we will regard the agents as 'attentive,' and assume $\varsigma^{\pm} = 1$.

3.3.1 *Characterisation of Expected Profit*

We recall from Sect. 3.2 that, just before the auction at time t, the agent j observes the value of his signal and works out his effective information $\bar{\sigma}(\xi_t^j)$ before he makes a judgement of the asset's value. Assuming $X \in \{X^l, X^h\}$ and, again, $|q_t^j| \in \{0, 1\}$, the expected (ex-ante) profit of agent j from his possible trade at time t can be decomposed as follows:

$$\mathbb{E}_t^j\left[\Pi_t^j\right] = P_t^j(\xi_c)\left|X^{l,h} - \mathbb{E}_t^j\left[S_t^*|\xi_c\right]\right| - P_t^j(\xi_e)\left|X^{l,h} - \mathbb{E}_t^j\left[S_t^*|\xi_e\right]\right| \tag{3.15}$$

with $P_t^j(\xi_c)$ and $P_t^j(\xi_e)$ being the chances of agent j getting correct and erroneous signals, i.e., ξ_c and ξ_e, at time t, respectively. And, again, $\mathbb{E}_t[\cdot] = \mathbb{E}[\cdot|\bar{\sigma}(\xi_t)]$. More formally,

$$\begin{aligned} P_t^j(\xi_c) &= P_t^j(H)P_t^j(\xi_c|H) + P_t^j(L)P^j(\xi_c|L) \qquad (3.16) \\ &= P_t^j(H)P^j(S_t^j > S_t^*|H) + P_t^j(L)P^j(S_t^j < S_t^*|L), \\ P_t^j(\xi_e) &= P_t^j(H)P_t^j(\xi_e|H) + P_t^j(L)P^j(\xi_e|L) \qquad (3.17) \\ &= P_t^j(H)P^j(S_t^j < S_t^*|H) + P_t^j(L)P^j(S_t^j > S_t^*|L), \end{aligned}$$

where, again, H and L denote high- and low-type markets in the sense of [18], and

$$
\begin{aligned}
(\xi_c|H) &:= (S_t^j > S_t^*|H), \quad (\xi_c|L) := (S_t^j < S_t^*|L) \\
(\xi_e|H) &:= (S_t^j < S_t^*|H), \quad (\xi_e|L) := (S_t^j > S_t^*|L).
\end{aligned}
\tag{3.18}
$$

Then, Eq. (3.15) can be written more explicitly as follows:

$$
\begin{aligned}
\mathbb{E}_t^j\left[\Pi_t^j\right] = P_t^j(H) & \left(P_t^j(\xi_c|H)\left(X^h - \mathbb{E}_t^j\left[S_t^*|H,\xi_c\right]\right)\right. \\
& \left.-P_t^j(\xi_e|H)\left(X^h - \mathbb{E}_t^j\left[S_t^*|H,\xi_e\right]\right)\right) + \\
& P_t^j(L)\left(P_t^j(\xi_c|L)\left(\mathbb{E}_t^j\left[S_t^*|L,\xi_c\right] - X^l\right)\right. \\
& \left.-P_t^j(\xi_e|L)\left(\mathbb{E}_t^j\left[S_t^*|L,\xi_e\right] - X^l\right)\right),
\end{aligned}
\tag{3.19}
$$

where

$$
\begin{aligned}
P_t^j(H) = P_t^j(X^h) &= \frac{p_h e^{\kappa_t f(t,\sigma_j,\xi^j,x_h)}}{\sum_{k\in\{h,l\}} p_k e^{\kappa_t f(t,\sigma_j,\xi^j,x_k)}}, \\
P_t^j(L) = P_t^j(X^l) &= \frac{p_l e^{\kappa_t f(t,\sigma_j,\xi^j,x_l)}}{\sum_{k\in\{h,l\}} p_k e^{\kappa_t f(t,\sigma_j,\xi^j,x_k)}},
\end{aligned}
\tag{3.20}
$$

with $f(t,\sigma,\xi,x) := \sigma\xi_t x - (1/2)\sigma^2 x^2 t$. When the payoff, i.e., $\phi(X) = X$, is continuous, however, Eq. (3.19) implies

$$
\begin{aligned}
\mathbb{E}_t^j\left[\Pi_t^j\right] = P_t^j(H) & \left(\int_{\mathbb{X}^h}\left(P_t^j(\xi_c|H,x)\left(x - \mathbb{E}_t^j\left[S_t^*(x)|H,\xi_c\right]\right)\right.\right. \\
& \left.\left.-P_t^j(\xi_e|H,x)\left(x - \mathbb{E}_t^j\left[S_t^*(x)|H,\xi_e\right]\right)\right)\pi_t^{j+}(x)\mathrm{d}x\right) + \\
& P_t^j(L)\left(\int_{\mathbb{X}^l}\left(P_t^j(\xi_c|L,x)\left(\mathbb{E}_t^j\left[S_t^*(x)|L,\xi_c)\right] - x\right)\right.\right. \\
& \left.\left.-P_t^j(\xi_e|L,x)\left(\mathbb{E}_t^j\left[S_t^*(x)|L,\xi_e\right] - x\right)\right)\pi_t^{j-}(x)\mathrm{d}x\right),
\end{aligned}
\tag{3.21}
$$

where $\mathbb{X}^h = (S_0^*, X_{\max})$ and $\mathbb{X}^l = (X_{\min}, S_0^*)$; π_t^{j+} and π_t^{j-} are normalised posteriors for high- and low-type markets, respectively; and, at this time,

$$
P_t^j(H) := \int_{\mathbb{X}^h}\pi_t^j(x)\mathrm{d}x, \text{ and } P_t^j(L) := \int_{\mathbb{X}^l}\pi_t^j(x)\mathrm{d}x.
\tag{3.22}
$$

The notation $S(x)$ is used to denote $\mathbb{E}[X|\xi(x)]$, i.e., the signal-based price of the agent when the actual signal is pinned to the value x. In a nutshell, expected profit of the agent is decomposed, through Eqs. (3.19) and (3.21), into two components, i.e., whether the agent's signal is pointing at the right (wrong) trade direction and, in that case, what the expected profit (loss) would be.

3.3.1.1 Trading Signal Quality: Digital Dividend

Assume, without loss of generality, that $X \in \{x_0, x_1\}$, with $x_0 = 0$, $x_1 > 0$ and the prior knowledge of the pair (p_0, p_1).[6] Let the true value of X be x_1. Equation (3.19) implies

$$\begin{aligned}\mathbb{E}_t^j\left[\Pi_t^j\right] = P_t^j(x_1)\Big(P_t^j(\xi_c|x_1)\left(x_1 - \mathbb{E}_t^j\left[S_t^*(x_1)|\xi_c\right]\right)\\ -P_t^j(\xi_e|x_1)\left(x_1 - \mathbb{E}_t^j\left[S_t^*(x_1)|\xi_e\right]\right)\Big) +\\ P_t^j(x_0)\Big(P_t^j(\xi_c|x_0)\left(\mathbb{E}_t^j\left[S_t^*(x_0)|\xi_c\right] - x_0\right)\\ -P_t^j(\xi_e|x_0)\left(\mathbb{E}_t^j\left[S_t^*(x_0)|\xi_e\right] - x_0\right)\Big).\end{aligned} \tag{3.23}$$

We can calculate the likelihoods of receiving a correct trade signal for agent 1 when $s_t = 0$ (i.e., no trade until t) in high- and low-type markets, respectively, as follows:

$$\begin{aligned}P_t^1(\xi_c|x_1) &= P\left(S_t^1(x_1) > S_t^*(x_1)\right) = P\left(S_t^1(x_1)/2 > S_t^2(x_1)/2\right)\\ &= P\left(\frac{\sum_{k=1,2} x_k p_k e^{\kappa_t f(t,\sigma_1,\xi^1,x_k)}}{\sum_{k=1,2} p_k e^{\kappa_t f(t,\sigma_1,\xi^1,x_k)}} > \frac{\sum_{k=1,2} x_k p_k e^{\kappa_t f(t,\sigma_2,\xi^2,x_k)}}{\sum_{k=1,2} p_k e^{\kappa_t f(t,\sigma_2,\xi^2,x_k)}}\right).\end{aligned} \tag{3.24}$$

A straightforward calculation yields

$$\begin{aligned}P_t^1(\xi_c|x_1) &= P\Big(x_1 p_1 e^{\kappa_t f(t,\sigma_1,\xi^1,x_1)} / \left(p_0 e^{\kappa_t f(t,\sigma_1,\xi^1,0)} + p_1 e^{\kappa_t f(t,\sigma_1,\xi^1,x_1)}\right)\\ &\quad > x_1 p_1 e^{\kappa_t f(t,\sigma_2,\xi^2,x_1)} / \left(p_0 e^{\kappa_t f(t,\sigma_2,\xi^2,0)} + p_1 e^{\kappa_t f(t,\sigma_2,\xi^2,x_1)}\right)\Big)\\ &= P\left(e^{\kappa_t f(t,\sigma_1,\xi^1,x_1)+\kappa_t f(t,\sigma_2,\xi^2,0)} > e^{\kappa_t f(t,\sigma_2,\xi^2,x_1)+\kappa_t f(t,\sigma_1,\xi^1,0)}\right)\end{aligned}$$

[6]Note that any binary payoff structure $X \in \{x_0, x_1\}$, $x_1 > x_0$ can be simplified as $\{0, x_1 - x_0\}$, a property which will simplify our calculations.

$$
\begin{aligned}
&= P\left(e^{\kappa_t f(t,\sigma_1,\xi^1,x_1)} > e^{\kappa_t f(t,\sigma_2,\xi^2,x_1)}\right) \quad (\text{note } f(t,\sigma,\xi,0)=0)\\
&= P\left(\kappa_t f(t,\sigma_1,\xi^1,x_1) > \kappa_t f(t,\sigma_2,\xi^2,x_1)\right)\\
&= P\left(\sigma_1\kappa_t\xi_t^1 x_1 - \frac{1}{2}\sigma_1^2\kappa_t x_1^2 t > \sigma_2\kappa_t\xi_t^2 x_1 - \frac{1}{2}\sigma_2^2 x_1^2 t\kappa_t\right)\\
&= P\left(\sigma_1\kappa_t\xi_t^1 - \sigma_2\kappa_t\xi_t^2 > \frac{1}{2}x_1 t\kappa_t\left(\sigma_1^2-\sigma_2^2\right)\right)\\
&= P\left(\sigma_1\kappa_t\left(z_1\sqrt{t\kappa_t^{-1}} + \sigma_1 x_1 t\right) - \sigma_2\kappa_t\left(z_2\sqrt{t\kappa_t^{-1}} + \sigma_2 x_1 t\right)\right.\\
&\qquad \left. > \frac{1}{2}x_1 t\kappa_t\left(\sigma_1^2-\sigma_2^2\right)\right)\\
&= P\left(\sigma_1\sqrt{t\kappa_t}z_1 - \sigma_2\sqrt{t\kappa_t}z_2 > \frac{1}{2}x_1 t\kappa_t\left(\sigma_2^2-\sigma_1^2\right)\right)\\
&= \Theta\left(\frac{1}{2}\frac{x_1 t\kappa_t\left(\sigma_1^2-\sigma_2^2\right)}{\left(\sigma_1^2+\sigma_2^2\right)^{1/2}\left(t\kappa_t\right)^{1/2}}\right) \quad (t<T), \qquad (3.25)
\end{aligned}
$$

where, again, $\Theta(\cdot)$ is the standard normal cumulative distribution function.[7] The last line simply follows from $z_1 \perp\!\!\!\perp z_2$ and $z_{1,2} \sim \mathcal{N}(0,1)$. Similarly, by arranging the last three lines of Eq. (3.25) and changing the direction of inequality from $>$ to $<$, we can indeed verify that

$$P_t^1(\xi_c|x_0) = P_t^1(\xi_c|x_1) \tag{3.26}$$

and, moreover, by virtue of convex combination in Eq. (3.16), that

$$P_t^1(\xi_c) = P_t^1(\xi_c|x_0) = P_t^1(\xi_c|x_1). \tag{3.27}$$

Equations (3.27) and (3.25) then directly imply

$$P_t^2(\xi_c) = 1 - P_t^1(\xi_c) = P_t^1(\xi_e). \tag{3.28}$$

Thus, the chances of agent j getting a correct (or erroneous) signal are the same no matter if the market is bullish or bearish, and one agent's success is the other one's failure, as expected. We now generalise Eq. (3.25) to the case where agents

[7] Note that, in Eq. (3.25), we inherently employ the basic relation $P(x > b-a) = 1 - \Theta(b-a) = \Theta(a-b)$.

did already exchange their information through trading, i.e. $s_t > 0$. Setting $s_t = s$, Eq. (3.25) can be rearranged as

$$\begin{aligned}
P_t^1(\xi_c|x_1, s) &= P\Big(x_1 p_1 e^{\kappa_t f(t,\sigma_1,\xi^1,x_1)+\kappa_s f(s,\sigma_2,\xi^2,x_1)} / \Big(p_0 + p_1 e^{\kappa_t f(t,\sigma_1,\xi^1,x_1)+\kappa_s f(s,\sigma_2,\xi^2,x_1)}\Big) \\
&\quad > x_1 p_1 e^{\kappa_t f(t,\sigma_2,\xi^2,x_1)+\kappa_s f(s,\sigma_1,\xi^1,x_1)} / \Big(p_0 + p_1 e^{\kappa_t f(t,\sigma_2,\xi^2,x_1)+\kappa_s f(s,\sigma_1,\xi^1,x_1)}\Big)\Big) \\
&= P\Big(e^{\kappa_t f(t,\sigma_1,\xi^1,x_1)+\kappa_s f(s,\sigma_2,\xi^2,x_1)} > e^{\kappa_t f(t,\sigma_2,\xi^2,x_1)+\kappa_s f(s,\sigma_1,\xi^1,x_1)}\Big) \\
&= P\big(\kappa_t f(t,\sigma_1,\xi^1,x_1) + \kappa_s f(s,\sigma_2,\xi^2,x_1) \\
&\quad > \kappa_t f(t,\sigma_2,\xi^2,x_1) + \kappa_s f(s,\sigma_1,\xi^1,x_1)\big) \\
&= P\Big(\sigma_1\kappa_t\xi_t^1 - \frac{1}{2}\sigma_1^2 x_1 t\kappa_t - \Big(\sigma_1\kappa_s\xi_s^1 - \frac{1}{2}\sigma_1^2 x_1 s\kappa_s\Big) \\
&\quad > \sigma_2\kappa_t\xi_t^2 - \frac{1}{2}\sigma_2^2 x_1 t\kappa_t - \Big(\sigma_2\kappa_s\xi_s^2 - \frac{1}{2}\sigma_2^2 x_1 s\kappa_s\Big)\Big) \\
&= P\Big(\sigma_1\left(\kappa_t\xi_t^1 - \kappa_s\xi_s^1\right) - \frac{1}{2}\sigma_1^2 x_1\left(t\kappa_t - s\kappa_s\right) \\
&\quad > \sigma_2\left(\kappa_t\xi_t^2 - \kappa_s\xi_s^2\right) - \frac{1}{2}\sigma_2^2 x_1\left(t\kappa_t - s\kappa_s\right)\Big) \\
&= P\big(\sigma_1\left(\kappa_t\left(\sigma_1 t x_1 + \beta_t^1\right) - \kappa_s\left(\sigma_1 s x_1 + \beta_s^1\right)\right) \\
&\quad - \sigma_2\left(\kappa_t\left(\sigma_2 t x_1 + \beta_t^2\right) - \kappa_s\left(\sigma_2 s x_1 + \beta_s^2\right)\right) \\
&\quad > \frac{1}{2} x_1\left(t\kappa_t - s\kappa_s\right)\left(\sigma_1^2 - \sigma_2^2\right)\Big) \\
&= P\Big(\sigma_1\left(\kappa_t\beta_t^1 - \kappa_s\beta_s^1\right) - \sigma_2\left(\kappa_t\beta_t^2 - \kappa_s\beta_s^2\right) > \frac{1}{2} x_1\left(t\kappa_t - s\kappa_s\right)\left(\sigma_2^2 - \sigma_1^2\right)\Big) \\
&= P\Big(\left(\sigma_1^2\kappa_t^2 t/\kappa_t - \sigma_1^2\kappa_s^2 s/\kappa_s\right)^{\frac{1}{2}} z_1 - \left(\sigma_2^2\kappa_t^2 t/\kappa_t - \sigma_2^2\kappa_s^2 s/\kappa_s\right)^{\frac{1}{2}} z_2 \\
&\quad > \frac{1}{2} x_1\left(t\kappa_t - s\kappa_s\right)\left(\sigma_2^2 - \sigma_1^2\right)\Big) \\
&= \Theta\left(\frac{1}{2}\frac{x_1(t\kappa_t - s\kappa_s)\left(\sigma_1^2 - \sigma_2^2\right)}{\left(\sigma_1^2 + \sigma_2^2\right)^{1/2}(t\kappa_t - s\kappa_s)^{1/2}}\right) \quad (s, t < T),
\end{aligned} \tag{3.29}$$

which, again, implies

- $P_t^1(\xi_c|x_0, s) = P_t^1(\xi_c|x_1, s)$,
- $P_t^1(\xi_c|s) = P_t^1(\xi_c|x_1, s) = P_t^1(\xi_c|x_0, s)$, and
- $P_t^2(\xi_c|s) = 1 - P_t^1(\xi_c|s) = P_t^1(\xi_e|s)$.

Note that neither $P_t^1(\xi_c|x_1)$ in Eq. (3.25) nor $P_t^1(\xi_c|x_1, s)$ in Eq. (3.29) is a function of the value of agent j's specific information at time t, i.e., $\bar{\sigma}(\xi_t^j)$, but rather depends only on the differential between information flow speeds, σ_1 and σ_2 (or, how agents perceive it), and the spread of X.

Equations (3.25) and (3.29) indeed reveal a number of intuitive properties, which agree with the analyses in Chap. 2 and earlier in this chapter, such as: (i) the larger the differential $|\sigma_1 - \sigma_2|$, the more likely the agent with superior signal will get a correct signal (ξ_c), (ii) with $|\sigma_1 - \sigma_2|$ given, the agent with superior signal will prefer more uncertainty (i.e., greater spread for X) to less uncertainty (i.e., smaller spread for X), and (iii) with $|\sigma_1 - \sigma_2|$ and spread of X given, refraining from a trade will always result in greater chances of getting a correct signal (although there will be a cost to refraining).

To complete the case where the contract pays a binary dividend, we state, by virtue of Eq. (3.30), the expected 'profit-to-go' of agent j at time t:

$$\sum_{u=t}^{T} \mathbb{E}_t^j\left[\Pi_{uT}^j\right] = \sum_{u=t}^{T} P_t^j(x_1)\left(P_u^j(\xi_c|x_1, s_u)\left(x_1 - \mathbb{E}_t^j\left[S_u^*(x_1)|\xi_c\right]\right)\right.$$
$$\left. - P_u^j(\xi_e|x_1)\left(x_1 - \mathbb{E}_t^j\left[S_u^*(x_1)|\xi_e\right]\right)\right) +$$
$$\sum_{u=t}^{T} P_t^j(x_0)\left(P_u^j(\xi_c|x_0)\left(\mathbb{E}_t^j\left[S_u^*(x_0)|\xi_c\right] - x_0\right)\right.$$
$$\left. - P_u^j(\xi_e|x_0)\left(\mathbb{E}_t^j\left[S_u^*(x_0)|\xi_e\right] - x_0\right)\right). \tag{3.30}$$

Note that we preserve the subscript t for p^j and $\mathbb{E}^j$ as they will be inferred based on the effective information at time t, i.e., $\bar{\sigma}(\xi_t^j)$. Below we generalise the above results to the case where X has a continuous distribution.

3.3.1.2 Trading Signal Quality: Gaussian Dividend

We first redefine the likelihoods of high- and low-type markets (see Eq. (3.22)):

$$P_t^j(x^+) = \int_{\mathbb{X}^+} \pi_t^j(x)\mathrm{d}x, \quad P_t^j(x^-) = \int_{\mathbb{X}^-} \pi_t^j(x)\mathrm{d}x, \tag{3.31}$$

where $\mathbb{X}^+ = (0, \infty)$ and $\mathbb{X}^- = (-\infty, 0)$. By virtue of Eq. (2.42) of Sect. 2.2.1, where we define the signal-based price when asset pays Gaussian dividends, the chances for agent 1 having right trade signals in high- and low-type markets can be found, in the same manner as Eq. (3.21), as follows.[8] Let $X = x$, $x > 0$,

$$\begin{aligned}
P_t^1(\xi_c|x^+) &= \int_{\mathbb{X}+} P^1(S_t^1(x) > S_t^*(x))\pi_t^+(x)\mathrm{d}x = \int_{\mathbb{X}+} P^1(S_t^1(x)/2 > S_t^2(x)/2)\pi_t^{1+}(x)\mathrm{d}x \\
&= \int_{\mathbb{X}+} P\left(\frac{1}{2}\frac{\sigma_1\kappa_t}{\sigma_1^2 t\kappa_t+1}\xi_t^1(x) - \frac{1}{2}\frac{\sigma_2\kappa_t}{\sigma_2^2 t\kappa_t+1}\xi_t^2(x) > 0\right)\pi_t^{1+}(x)\mathrm{d}x \\
&= \int_{\mathbb{X}+} P\left(\frac{1}{2}\frac{\sigma_1\kappa_t}{\sigma_1^2 t\kappa_t+1}(\sigma_1 tx + \beta_t^1) - \frac{1}{2}\frac{\sigma_2\kappa_t}{\sigma_2^2 t\kappa_t+1}(\sigma_2 tx + \beta_t^2) > 0\right)\pi_t^{1+}(x)\mathrm{d}x \\
&= \int_{\mathbb{X}+} P\left(\frac{1}{2}\frac{\sigma_1\kappa_t}{\sigma_1^2 t\kappa_t+1}\sqrt{t/\kappa_t}z_1 - \frac{1}{2}\frac{\sigma_2\kappa_t}{\sigma_2^2 t\kappa_t+1}\sqrt{t/\kappa_t}z_2 \quad (z_1 \perp\!\!\!\perp z_2,\, z_{1,2}\sim\mathcal{N}(0,1))\right. \\
&\qquad \left. > \frac{1}{2}xt\kappa_t\left(\frac{\sigma_2^2}{\sigma_2^2 t\kappa_t+1} - \frac{\sigma_1^2}{\sigma_1^2 t\kappa_t+1}\right)\right)\pi_t^{1+}(x)\mathrm{d}x \\
&= \int_{\mathbb{X}+} \Theta\left(\frac{\frac{1}{2}xt\kappa_t\left(\frac{\sigma_1^2}{\sigma_2^2 t\kappa_t+1} - \frac{\sigma_2^2}{\sigma_1^2 t\kappa_t+1}\right)}{\frac{1}{2}\sqrt{t\kappa_t}\sqrt{\left(\frac{\sigma_1}{\sigma_1^2 t\kappa_t+1}\right)^2 + \left(\frac{\sigma_2}{\sigma_2^2 t\kappa_t+1}\right)^2}}\right)\pi_t^{1+}(x)\mathrm{d}x \\
&= \int_{\mathbb{X}+} \Theta\left(-\frac{a_t}{b_t}x\right)\pi_t^{1+}(x)\mathrm{d}x \quad (a_t, b_t > 0, x > 0),
\end{aligned} \tag{3.32}$$

where π_t^+ is the normalised density,

$$a_t = \frac{1}{2}t\kappa_t\left(\frac{\sigma_2^2}{\sigma_2^2 t\kappa_t+1} - \frac{\sigma_1^2}{\sigma_1^2 t\kappa_t+1}\right) \tag{3.33}$$

and

$$b_t = \frac{1}{2}\sqrt{t\kappa_t}\left(\left(\frac{\sigma_1}{\sigma_1^2 t\kappa_t+1}\right)^2 + \left(\frac{\sigma_2}{\sigma_2^2 t\kappa_t+1}\right)^2\right)^{1/2}. \tag{3.34}$$

Accordingly, for a low-type market, it can be shown that

$$\begin{aligned}
P_t^1(\xi_c|x^-) &= \int_{\mathbb{X}-} \Theta\left(\frac{a_t}{b_t}x\right)\pi_t^{1-}(x)\mathrm{d}x \quad (a_t, b_t > 0, x_t < 0) \\
&= \int_{\mathbb{X}+} \Theta\left(-\frac{a_t}{b_t}x\right)\pi_t^{1+}(x)\mathrm{d}x \quad (a_t, b_t > 0, x > 0) \\
&= P_t^1(\xi_c|x^+).
\end{aligned} \tag{3.35}$$

[8]We use the property that the price process S_t is Gaussian when X has a Gaussian terminal distribution.

If $s_t = s$ is non-zero, in which case Eq. (2.42) takes the form

$$S_t^1 = \mathbb{1}_{\{t<T\}} e^{-r(T-t)} \frac{\sigma_1 \kappa_t \xi_t^1 + \sigma_2 \kappa_s \xi_s^2}{\sigma_1^2 \kappa_t t + \sigma_2^2 \kappa_s s + 1}, \tag{3.36}$$

we find $P_t^1(\xi_c|x^+, s)$ as

$$\begin{aligned}
P_t^1(\xi_c|x^+, s) &= \int_{\mathbb{X}+} P^1(S_t^1(x) > S_t^*(x)|s)\bar{\pi}_t^{1+}(x)\mathrm{d}x \\
&= \int_{\mathbb{X}+} P^1(S_t^1(x)/2 > S_t^2(x)/2|s)\bar{\pi}_t^{1+}(x)\mathrm{d}x \\
&= \int_{\mathbb{X}+} P\left(\frac{1}{2}\frac{\sigma_1\kappa_t\xi_t^1 + \sigma_2\kappa_s\xi_s^2}{\sigma_1^2 t\kappa_t + \sigma_2^2 s\kappa_s + 1} - \frac{1}{2}\frac{\sigma_2\kappa_t\xi_t^2 + \sigma_1\kappa_s\xi_s^1}{\sigma_2^2 t\kappa_t + \sigma_1^2 s\kappa_s + 1} > 0\right)\bar{\pi}_t^{1+}(x)\mathrm{d}x \\
&= \int_{\mathbb{X}+} P\left(\frac{1}{2}\frac{\sigma_1\kappa_t(\sigma_1 tx + \beta_t^1)}{\sigma_1^2 t\kappa_t + \sigma_2^2 s\kappa_s + 1} - \frac{1}{2}\frac{\sigma_1\kappa_s(\sigma_1 sx + \beta_s^1)}{\sigma_1^2 s\kappa_s + \sigma_2^2 t\kappa_t + 1}\right. \\
&\quad \left. - \left(\frac{1}{2}\frac{\sigma_2\kappa_t(\sigma_2 tx + \beta_t^2)}{\sigma_2^2 t\kappa_t + \sigma_1^2 s\kappa_s + 1} - \frac{1}{2}\frac{\sigma_2\kappa_s(\sigma_2 sx + \beta_s^2)}{\sigma_2^2 s\kappa_s + \sigma_1^2 t\kappa_t + 1}\right) > 0\right)\bar{\pi}_t^{1+}(x)\mathrm{d}x \\
&= \int_{\mathbb{X}+} P\left\{\frac{1}{2}\left[\left(\frac{\sigma_1\kappa_t}{\sigma_1^2 t\kappa_t + \sigma_2^2 s\kappa_s + 1}\right)^2 (t/\kappa_t) + \left(\frac{\sigma_1\kappa_s}{\sigma_1^2 s\kappa_s + \sigma_2^2 t\kappa_t + 1}\right)^2 (s/\kappa_s)\right.\right. \\
&\quad \left. -2\left(\frac{\sigma_1\kappa_t}{\sigma_1^2 t\kappa_t + \sigma_2^2 s\kappa_s + 1}\right)\left(\frac{\sigma_1\kappa_s}{\sigma_1^2 s\kappa_s + \sigma_2^2 t\kappa_t + 1}\right)(s/\kappa_t)\right]^{\frac{1}{2}} z_1 \\
&\quad -\frac{1}{2}\left[\left(\frac{\sigma_2\kappa_t}{\sigma_2^2 t\kappa_t + \sigma_1^2 s\kappa_s + 1}\right)^2 (t/\kappa_t) + \left(\frac{\sigma_2\kappa_s}{\sigma_2^2 s\kappa_s + \sigma_1^2 t\kappa_t + 1}\right)^2 (s/\kappa_s)\right. \\
&\quad \left. -2\left(\frac{\sigma_2\kappa_t}{\sigma_2^2 t\kappa_t + \sigma_1^2 s\kappa_s + 1}\right)\left(\frac{\sigma_2\kappa_s}{\sigma_2^2 s\kappa_s + \sigma_1^2 t\kappa_t + 1}\right)(s/\kappa_t)\right]^{\frac{1}{2}} z_2 \\
&\quad > \frac{1}{2}\left(\frac{\sigma_2^2 t\kappa_t x}{\sigma_2^2 t\kappa_t + \sigma_1^2 s\kappa_s + 1} - \frac{\sigma_2^2 s\kappa_s x}{\sigma_2^2 s\kappa_s + \sigma_1^2 t\kappa_t + 1}\right) \\
&\quad \left. -\frac{1}{2}\left(\frac{\sigma_1^2 t\kappa_t x}{\sigma_1^2 t\kappa_t + \sigma_2^2 s\kappa_s + 1} - \frac{\sigma_1^2 s\kappa_s x}{\sigma_1^2 s\kappa_s + \sigma_2^2 t\kappa_t + 1}\right)\right\}\bar{\pi}_t^{1+}(x)\mathrm{d}x \\
&= \int_{\mathbb{X}+} \Theta\left(-\frac{a_t^s}{b_t^s}x\right)\bar{\pi}_t^{1+}(x)\mathrm{d}x \quad (a_t^s, b_t^s, x > 0),
\end{aligned} \tag{3.37}$$

where, again, $z_{1,2}$ are independently $\mathcal{N}(0,1)$, $\bar{\pi}^{+}$ is the normalised effective posterior density as given in Eq. (3.9),

$$a_t^s = \frac{1}{2}\left(\frac{\sigma_2^2 t\kappa_t}{\sigma_2^2 t\kappa_t + \sigma_1^2 s\kappa_s + 1} - \frac{\sigma_2^2 s\kappa_s}{\sigma_2^2 s\kappa_s + \sigma_1^2 t\kappa_t + 1}\right)$$
$$-\frac{1}{2}\left(\frac{\sigma_1^2 t\kappa_t}{\sigma_1^2 t\kappa_t + \sigma_2^2 s\kappa_s + 1} - \frac{\sigma_1^2 s\kappa_s}{\sigma_1^2 s\kappa_s + \sigma_2^2 t\kappa_t + 1}\right), \tag{3.38}$$

$$b_t^s = \frac{1}{2}\left[\left(\frac{\sigma_1\kappa_t}{\sigma_1^2 t\kappa_t + \sigma_2^2 s\kappa_s + 1}\right)^2 (t/\kappa_t) + \left(\frac{\sigma_1\kappa_s}{\sigma_1^2 s\kappa_s + \sigma_2^2 t\kappa_t + 1}\right)^2 (s/\kappa_s)\right.$$
$$-2\left(\frac{\sigma_1\kappa_t}{\sigma_1^2 t\kappa_t + \sigma_2^2 s\kappa_s + 1}\right)\left(\frac{\sigma_1\kappa_s}{\sigma_1^2 s\kappa_s + \sigma_2^2 t\kappa_t + 1}\right)(s/\kappa_t)$$
$$+\left(\frac{\sigma_2\kappa_t}{\sigma_2^2 t\kappa_t + \sigma_1^2 s\kappa_s + 1}\right)^2 (t/\kappa_t) + \left(\frac{\sigma_2\kappa_s}{\sigma_2^2 s\kappa_s + \sigma_1^2 t\kappa_t + 1}\right)^2 (s/\kappa_s)$$
$$\left.-2\left(\frac{\sigma_2\kappa_t}{\sigma_2^2 t\kappa_t + \sigma_1^2 s\kappa_s + 1}\right)\left(\frac{\sigma_2\kappa_s}{\sigma_2^2 s\kappa_s + \sigma_1^2 t\kappa_t + 1}\right)(s/\kappa_t)\right]^{\frac{1}{2}}. \tag{3.39}$$

Indeed, we can quickly verify that $a_t^s > 0$.[9] Moreover, similar to the digital payoff case, it directly follows from (3.37) that $P_t^1(\xi_c|x^+, s) = P_t^1(\xi_c|x^-, s)$, $p_t^1(\xi_c|s) = P_t^1(\xi_c|x^+, s) = P_t^1(\xi_c|x^-, s)$, and $p_t^2(\xi_c|s) = 1 - p_t^1(\xi_c|s) = p_t^1(\xi_e|s)$. Note in addition that, for $s = 0$, Eq. (3.37) simplifies to Eq. (3.32).

For a low-type market, similarly,

$$\begin{aligned} P_t^1(\xi_c|x^-, s) &= \int_{\mathbb{X}^-} \Theta\left(\frac{a_t^s}{b_t^s}x\right)\pi_t^-(x)\mathrm{d}x \quad (a_t^s, b_t^s > 0, x < 0) \\ &= \int_{\mathbb{X}^+} \Theta\left(-\frac{a_t^s}{b_t^s}x\right)\pi_t^+(x)\mathrm{d}x \quad (a_t^s, b_t^s > 0, x > 0) \\ &= P_t^1(\xi_c|x^+, s). \end{aligned} \tag{3.40}$$

Expected profit-to-go at time t can be inferred in a similar sense to Eq. (3.30).

As a final step to calculate the signal-based expected profit of then agent at time t (i.e., just before the auction), as given in Eq. (3.21), we now compute the expected transaction price in low- and high-type markets and with correct and erroneous

[9] As $x > y$ implies $x/(x+1) > y/(y+1)$.

signals. Note that

$$\begin{aligned}\mathbb{E}_t^1\left[S_t^*(x)\middle|x^+,\xi_c,s\right] &= 1/2\left(S_t^1(x)+\mathbb{E}_t^1\left[S_t^2(x)\middle|x^+,\xi_c,s\right]\right)\\ &= S_t^1(x)-\mathbb{E}\left[1/2\left(S_t^1(x)-S_t^2(x)\right)^+\middle|x^+,s\right],\end{aligned}\tag{3.41}$$

$$\begin{aligned}\mathbb{E}_t^1\left[S_t^*(x)\middle|x^+,\xi_e,s\right] &= 1/2\left(S_t^1(x)+\mathbb{E}_t^1\left[S_t^2(x)\middle|x^+,\xi_e,s\right]\right)\\ &= S_t^1(x)+\mathbb{E}\left[1/2\left(S_t^2(x)-S_t^1(x)\right)^+\middle|x^+,s\right],\end{aligned}\tag{3.42}$$

$$\begin{aligned}\mathbb{E}_t^1\left[S_t^*(x)\middle|x^-,\xi_c\right] &= 1/2\left(S_t^1(x)+\mathbb{E}_t^1\left[S_t^2(x)\middle|x^-,\xi_c,s\right]\right)\\ &= S_t^1(x)+\mathbb{E}\left[1/2\left(S_t^2(x)-S_t^1(x)\right)^+\middle|x^-,s\right]\end{aligned}\tag{3.43}$$

and

$$\begin{aligned}\mathbb{E}_t^1\left[S_t^*(x)\middle|x^-,\xi_e\right] &= 1/2\left(S_t^1(x)+\mathbb{E}_t^1\left[S_t^2(x)\middle|x^-,\xi_e,s\right]\right)\\ &= S_t^1(x)-\mathbb{E}\left[1/2\left(S_t^1(x)-S_t^2(x)\right)^+\middle|x^-,s\right].\end{aligned}\tag{3.44}$$

Notice that we dropped t and j from $\mathbb{E}_t^j$'s in Eqs. (3.41)–(3.44) since, by Eq. (3.37), when $(\xi_t^j)_{0\le t\le T}$ is pinned to a certain value x, the price differential is not conditional on the specific value of agent j's signal at time t, but rather a function of σ_1, σ_2, t and s. Thus, all one needs to do (so as to compute the expected transaction price under different market situations and trading signal quality) is to work out the expected value of the 'absolute price differential' under each circumstance. To that end, we can infer from Eq. (3.37) that

$$\begin{aligned}\left(1/2\left(S_t^2(x)-S_t^1(x)\right)\middle|x^+,s\right) &\sim \mathcal{N}(a_t^s x, b_t^s),\\ \left(1/2\left(S_t^2(x)-S_t^1(x)\right)\middle|x^-,s\right) &\sim \mathcal{N}(a_t^s x, b_t^s),\\ \left(1/2\left(S_t^1(x)-S_t^2(x)\right)\middle|x^+,s\right) &\sim \mathcal{N}(-a_t^s x, b_t^s),\\ \left(1/2\left(S_t^1(x)-S_t^2(x)\right)\middle|x^-,s\right) &\sim \mathcal{N}(-a_t^s x, b_t^s),\end{aligned}$$

where $a_t^s, b_t^s > 0$ are as given above. As a result,

$$\begin{aligned}\mathbb{E}\left[1/2\left(S_t^2(x)-S_t^1(x)\right)^+\middle|x^+,s\right] &= \begin{cases}\eta_1\frac{1}{\sqrt{2\pi}b_t^s}\int_0^\infty ye^{-\frac{1}{2}\frac{(y-a_t^s x)^2}{(b_t^s)^2}}\,\mathrm{d}y, & x>0,\\ \eta_1\frac{1}{\sqrt{2\pi}b_t^s}\int_0^\infty ye^{-\frac{1}{2}\frac{(y+a_t^s x)^2}{(b_t^s)^2}}\,\mathrm{d}y, & x<0,\end{cases}\\ &= \mathbb{E}\left[1/2\left(S_t^1(x)-S_t^2(x)\right)^+\middle|x^-,s\right],\end{aligned}\tag{3.45}$$

and

$$\mathbb{E}\left[1/2\left(S_t^1(x)-S_t^2(x)\right)^+\left|x^+,s\right.\right] = \begin{cases} \eta_2 \frac{1}{\sqrt{2\pi}b_t^s}\int_0^\infty ye^{-\frac{1}{2}\frac{(y+a_t^s x)^2}{(b_t^s)^2}}\mathrm{d}y,\ x>0,\\ \eta_2 \frac{1}{\sqrt{2\pi}b_t^s}\int_0^\infty ye^{-\frac{1}{2}\frac{(y-a_t^s x)^2}{(b_t^s)^2}}\mathrm{d}y,\ x<0, \end{cases}$$
$$= \mathbb{E}\left[1/2\left(S_t^2(x)-S_t^1(x)\right)^+\left|x^-,s\right.\right], \tag{3.46}$$

with density normalising factors η_1, η_2. Thus, under the Gaussian payoff scenario, we have derived explicit formulae for the two main sources of uncertainty involved in signal-based trade, namely, the likelihood of a signal's pointing at the right (wrong) trade direction, and the expected amount of profit (loss) given the signal was correct (erroneous). With inputs from Eqs. (3.37) and (3.41)–(3.44), Eq. (3.21) can now be written for agent 1 as

$$\begin{aligned} \mathbb{E}_t^1\left[\Pi_t^1\right] = P_t^1(x^+)&\left(\int_{\mathbb{X}^+}\left(\Theta\left(-\frac{a_t^s}{b_t^s}x\right)\eta_1\frac{1}{\sqrt{2\pi}b_t^s}\int_0^\infty ye^{-\frac{1}{2}\frac{(y+a_t^s x)^2}{(b_t^s)^2}}\mathrm{d}y\right.\right.\\ &\left.\left.-\Theta\left(\frac{a_t^s}{b_t^s}x\right)\eta_2\frac{1}{\sqrt{2\pi}b_t^s}\int_0^\infty ye^{-\frac{1}{2}\frac{(y-a_t^s x)^2}{(b_t^s)^2}}\mathrm{d}y\right)\bar{\pi}_t^{1+}(x)\mathrm{d}x\right)+\\ P_t^1(x^-)&\left(\int_{\mathbb{X}^-}\left(\Theta\left(\frac{a_t^s}{b_t^s}x\right)\eta_1\frac{1}{\sqrt{2\pi}b_t^s}\int_0^\infty ye^{-\frac{1}{2}\frac{(y-a_t^s x)^2}{(b_t^s)^2}}\mathrm{d}y\right.\right.\\ &\left.\left.-\Theta\left(-\frac{a_t^s}{b_t^s}x\right)\eta_2\frac{1}{\sqrt{2\pi}b_t^s}\int_0^\infty ye^{-\frac{1}{2}\frac{(y+a_t^s x)^2}{(b_t^s)^2}}\mathrm{d}y\right)\bar{\pi}_t^{1-}(x)\mathrm{d}x\right), \end{aligned} \tag{3.47}$$

where $\bar{\pi}^+$, $\bar{\pi}^-$ are, again, normalised effective posteriors for high- and low-type markets, with $\mathbb{X}^+=(0,\infty)$ and $\mathbb{X}^-=(-\infty,0)$. Equation (3.47) can also be written for agent 2 without much effort. Accordingly, agent j updates his trading procedure as follows: (1)—(1a) Same as in Sect. 3.2 above. (1b) Calculate $\mathbb{E}_t^j[\Pi_t^j]$ based on Eq. (3.47). (1c) Decide whether to quote or not to quote a price. If yes, proceed to next step. (2) Quote signal-based price (as $\varsigma^{\pm}=1$). (3) Same as in Sect. 3.2 above.

Now, equipped with the flexibility to shape his strategy $(q_t^j)_{0\le t\le T}$, $|q_t^j|\in\{0,1\}$, by timing his trades, agent j will need to develop an optimal 'online' trading rule (referring to (1c) above) that maximises his profits, in understanding of his marginal benefits and losses from seizing or skipping a trade opportunity.

3.3.2 Risk-Neutral Optimal Strategy

It is not difficult to see that $\mathbb{E}_t^1\left[\Pi_t^1\right]$ in Eq. (3.47) will always be negative when agent knows that his information is less superior. The top and bottom left panels of Fig. 3.8, in this regard, depicts the evolution in time of $\mathbb{E}_t^1\left[\Pi_t^1\right]$, based on almost all possible strategies and a sample path of ξ_t^1, when an agent believes that he is informationally less susceptible than his counterpart. Yet, the agent can minimise his chances of losing from a trade by keeping his information up-to-date through trading at "each" time step (i.e., the top edge of each shape in the left panel). We note that the marginal expected cost of refraining from trade for the less susceptible agent is always positive when the agent believes he is informationally less susceptible.

We therefore infer from Fig. 3.8 that a solution to the maximisation problem in Eq. (3.48) is unattainable from the perspective of a less informationally capable agent. The real-world implication of this is that market shutdowns may not occur in a real market setting because investors think their effective information are either constantly or temporarily superior to the market information. We, thus, turn our focus to the case where both agents believe their information source is characterised by a higher σ.

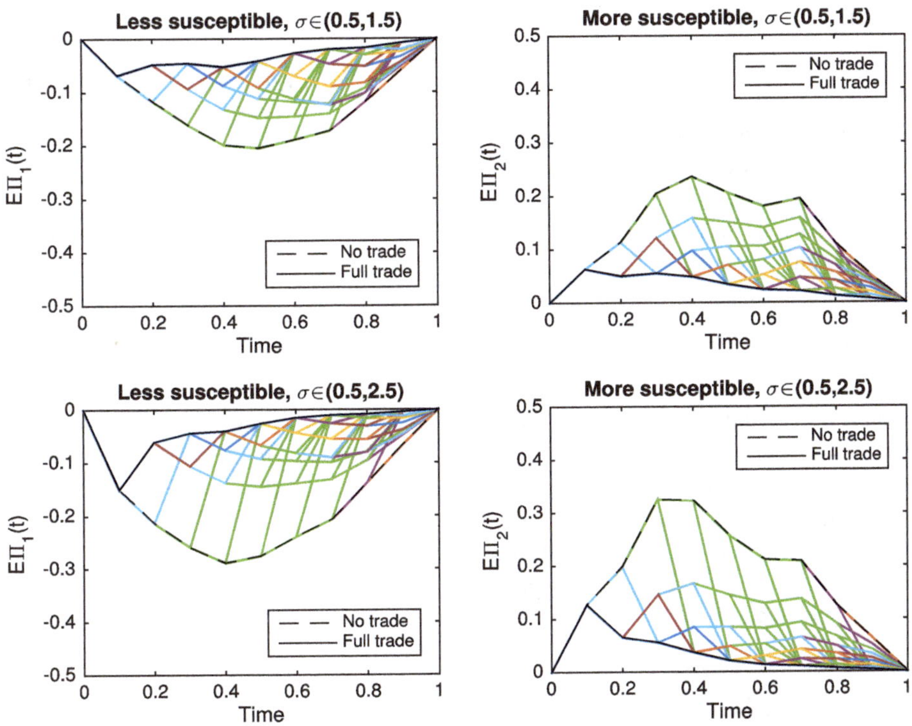

Fig. 3.8 Evolution of $\mathbb{E}_t^j[\Pi_t^j]$ as given in Eq. (3.47) for sample trajectories of ξ_t^1 and ξ_t^2 and all possible trading strategies. The dividend is assumed to be Gaussian

The top and bottom right panels of Fig. 3.8, on the other hand, shows the evolution in time of $\mathbb{E}_t^j\left[\Pi_t^j\right]$ for the agent who believes that he is informationally more susceptible. The strategy which results in the bottom edge these shapes on the right panel is unique, i.e., $|q_t^j| = 1\ \forall t$. However, there is no single strategy which can achieve the top edge of each shape, which is a combination of different strategies that result in the maximum expected potential at different time points.

We define the optimal strategy of an agent as the one which maximises his overall expected terminal profit from trading the contract based on his effective information $\bar{\sigma}(\xi_t^j)$, i.e., for agent 1,

$$\begin{aligned} \arg\max_{(q_t^1)} \quad & \sum_{t=0}^{T} \mathbb{E}_t^1\left[\Pi_t^1\right] = \sum_{t=0}^{T} \mathbb{E}\left[\Pi_t^1 | \bar{\sigma}(\xi_t^1)\right] \\ \text{s.t.} \quad & S_t^* = 1/2(S_t^1 + S_t^2), \\ & q_t = 0 \\ & \forall t, \end{aligned} \tag{3.48}$$

where $\bar{\sigma}(\xi_t^1)$ is same as in Eq. (3.8).

On an extra note, we remark that setting a mid-price, as in Eq. (3.2) and (3.48), is indeed equivalent to

$$\mathbb{E}\left[q_t^1(X - S_t^*)|\bar{\sigma}(\xi_t^1)\right] = \mathbb{E}\left[q_t^2(X - S_t^*)|\bar{\sigma}(\xi_t^2)\right]. \tag{3.49}$$

Thus, we can reinterpret the role of the central planner, in the context of this section, as 'to observe ξ_t^j's through price quotes and set the transaction price as the mid-price which equates the signal-based terminal profits of agents.'

Similar to [11], we can define the dynamic programming formulation of the agent j's problem given in Eq. (3.48) as follows:

$$V_t^j = \sup_{(q_t^j)} \left(\mathbb{E}_t^j\left[\Pi_t^j\right] + \mathbb{E}_t^j\left[V_{t+1}^j\right]\right), \tag{3.50}$$

where V_t^j is the value function. Note that $\mathbb{E}_t^j[\Pi_t^j]$ is implicitly determined by $(q_s^j)_{0\leq s<t}$, whereas $\mathbb{E}_t^j[V_{t+1}^j]$ by $(q_s^j)_{0\leq s\leq t}$. Therefore, at each auction, the agent will need to consider the marginal impact of his current strategy on $\mathbb{E}_t^j[V_{t+1}^j]$. The particular nature of the present model, however, does not allow us to employ a backward-induction technique that is similar to the one described in [11, 14].

Based on Eq. (3.50), we introduce the following real-time optimal trading strategy for agent j:

$$|q_t^j| = \begin{cases} 1, \text{ if } \ \mathbb{E}_t^j\left[\Pi_t^j\right] > 0, \\ \quad \text{and } \ \mathbb{E}_t^j\left[\Pi_t^j\right] + \mathbb{E}_t^j\left[\Pi_{t+1}^j\right]_{|q_t^j|=1} > \mathbb{E}_t^j\left[\Pi_{t+1}^j\right]_{q_t^j=0}, \\ 0, \text{ otherwise,} \end{cases} \tag{3.51}$$

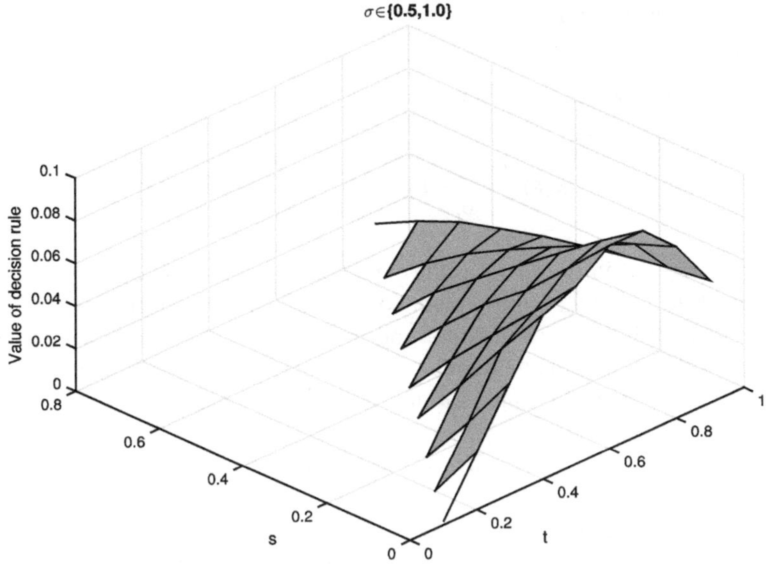

Fig. 3.9 Value of cost-adjusted expected gain from trade (averaged over a number of sample paths of ξ_t^j) for the more informationally susceptible agent based on Eq. (3.51) for all t and s_t

where the term

$$\mathbb{E}_t^j[\Pi_t^j] - \left(\mathbb{E}_t^j[\Pi_{t+1}^j]_{q_t^j=0} - \mathbb{E}_t^j[\Pi_{t+1}^j]_{|q_t^j|=1}\right) \tag{3.52}$$

can be seen as the immediate expected gain from trade adjusted for the cost of losing the informational advantage. Thus, the agent chooses to trade whenever his cost-adjusted expected gain from trade is strictly positive.[10]

Figure 3.9 plots the value of (3.52) (averaged over a number of sample paths of ξ_t^j) for the more informationally susceptible agent for each point t in the trading horizon and given each possible trading history s_t. The decision rule variable is positive for any possible past strategy characterised by the last time of trade, $q_{s_t}^j$, implying that the agent can maximise the sum of his expected terminal profits by trading at each time point t, thereby constantly incorporating his differential information into prices, i.e., $|q_t^j| = 1 \ \forall t \in [0, T]$.

We can indeed show that the optimality of this strategy is invariant to the path of ξ_t^j. Consider agent 2 who deems his signal superior ($\sigma_2 > \sigma_1$) and let the market be high-type (i.e., $x \in \mathbb{X}^+$). For given x, σ_1, σ_2, let's denote the corresponding integrand

[10] In other words, whenever an agent refrains from trade in expectation of greater future profits, he should refrain on the basis that he has to recover immediate cost of refraining.

in Eq. (3.47), rearranged for agent 2, by $H_2(t,s)$, where

$$H_2(t,s) = \Theta\left(\frac{a_t^s}{b_t^s}x\right)\eta_1 \frac{1}{\sqrt{2\pi}b_t^s}\int_0^\infty y e^{-\frac{1}{2}\frac{(y-a_t^s x)^2}{(b_t^s)^2}}\mathrm{d}y$$
$$-\Theta\left(-\frac{a_t^s}{b_t^s}x\right)\eta_1 \frac{1}{\sqrt{2\pi}b_t^s}\int_0^\infty y e^{-\frac{1}{2}\frac{(y+a_t^s x)^2}{(b_t^s)^2}}\mathrm{d}y. \quad (3.53)$$

Note that $H_2(t,s)$ in Eq. (3.53) is the expected profit of agent 2 at time t given the time of last trade, s, and a high-type payoff x, and it is not a function of ξ_t^2. Agent 2 version of Eq. (3.47) is, in fact, nothing but the sum of convex combinations of $H_2(t,s)$ and its low-market analogous $L_2(t,s)$ with respect to the effective posteriors $\bar{\pi}_t^{2+}$ and $\bar{\pi}_t^{2-}$, respectively. Thus, similar to the relation (3.52),

$$H_2(t,s) - [H_2(t+1,s) - H_2(t+1,t)] \quad (3.54)$$

can be seen as the signal-independent version of the cost-adjusted immediate gain from trade (for the agent who deems his signal superior), whose value is depicted in Fig. 3.10. It can be inferred from the figure, again, that it is optimal for the informationally more susceptible agent to trade continuously without accumulating his extra information.

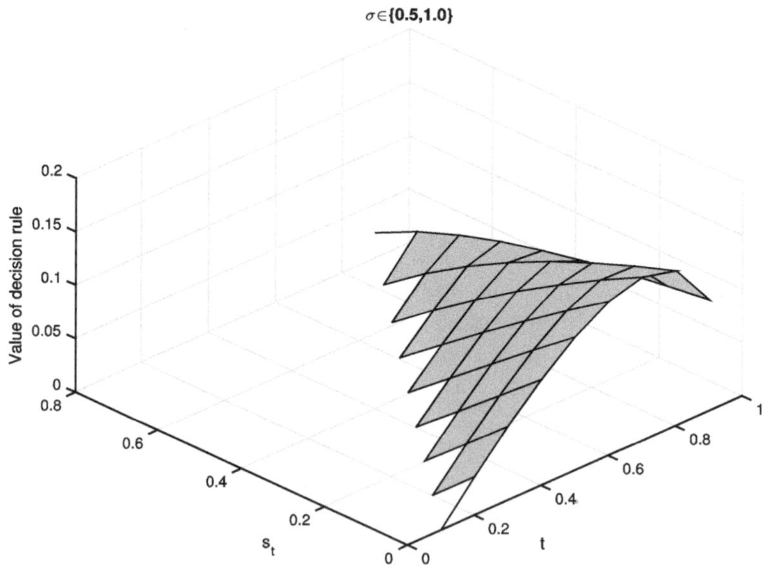

Fig. 3.10 Value of signal-independent cost-adjusted expected gain from trade for the more informationally susceptible agent based on Eq. (3.51) for all t and s_t

3.3.3 Extension to Risk-Adjusted Performance

In case agents are risk-adjusted expected profit (e.g., Sharpe ratio, [20]) maximisers "at the portfolio level", the objective function in Eq. (3.48) can simply be modified as

$$\arg\max_{(q_t^j)} \frac{\sum_{t=0}^{T} q_t^j \mathbb{E}_t^j\left[\Pi_t^j\right]}{\left(\sum_{t=0}^{T} (q_t^j)^2 \mathbb{V}_t^j\left(\Pi_t^j\right)\right)^{1/2}}. \tag{3.55}$$

We then write the conditional variance $\mathbb{V}_t^j(\Pi_t^j) = \mathbb{V}^j(\Pi_t^j|\bar{\sigma}(\xi_t^j))$ of the signal-based profit at time t, whose expectation is given in Eq. (3.47), using the "law of total variance",[11] as

$$\begin{aligned}\mathbb{V}_t^1(\Pi_t^1) = p_t^1(x^+)\Bigg(\int_{\mathbb{X}^+}\Bigg(\Theta\left(-\frac{a_t^s}{b_t^s}x\right)^2 \eta_1 \frac{1}{\sqrt{2\pi}b_t^s}\int_0^\infty y^2 e^{-\frac{1}{2}\frac{(y+a_t^s x)^2}{(b_t^s)^2}}\mathrm{d}y \\ -\Theta\left(\frac{a_t^s}{b_t^s}x\right)^2 \eta_2 \frac{1}{\sqrt{2\pi}b_t^s}\int_0^\infty y^2 e^{-\frac{1}{2}\frac{(y-a_t^s x)^2}{(b_t^s)^2}}\mathrm{d}y\Bigg)\bar{\pi}_t^{1+}(x)\mathrm{d}x\Bigg) + \\ p_t^1(x^-)\Bigg(\int_{\mathbb{X}^-}\Bigg(\Theta\left(\frac{a_t^s}{b_t^s}x\right)^2 \eta_1 \frac{1}{\sqrt{2\pi}b_t^s}\int_0^\infty y^2 e^{-\frac{1}{2}\frac{(y-a_t^s x)^2}{(b_t^s)^2}}\mathrm{d}y \\ -\Theta\left(-\frac{a_t^s}{b_t^s}x\right)^2 \eta_2 \frac{1}{\sqrt{2\pi}b_t^s}\int_0^\infty y^2 e^{-\frac{1}{2}\frac{(y+a_t^s x)^2}{(b_t^s)^2}}\mathrm{d}y\Bigg)\bar{\pi}_t^{1-}(x)\mathrm{d}x\Bigg) - \left(\mathbb{E}_t^1[\Pi_t^1]\right)^2.\end{aligned} \tag{3.56}$$

Figure 3.11 depicts the risk-adjusted version of Fig. 3.8 using the same signal sample as in the latter.

3.3.4 Extension to Risk-Averse Utility

The above setup can easily be generalised to the case where agents are 'characteristically' risk-averse and attach decreasing marginal utility to each extra unit of expected return due to the additional risk involved. In [6], the authors show

[11] $\mathbb{V}[\Pi|\xi] = \mathbb{E}[\mathbb{E}[\Pi^2|\xi, X]] - \mathbb{E}[\mathbb{E}[\Pi|\xi, X]]^2$.

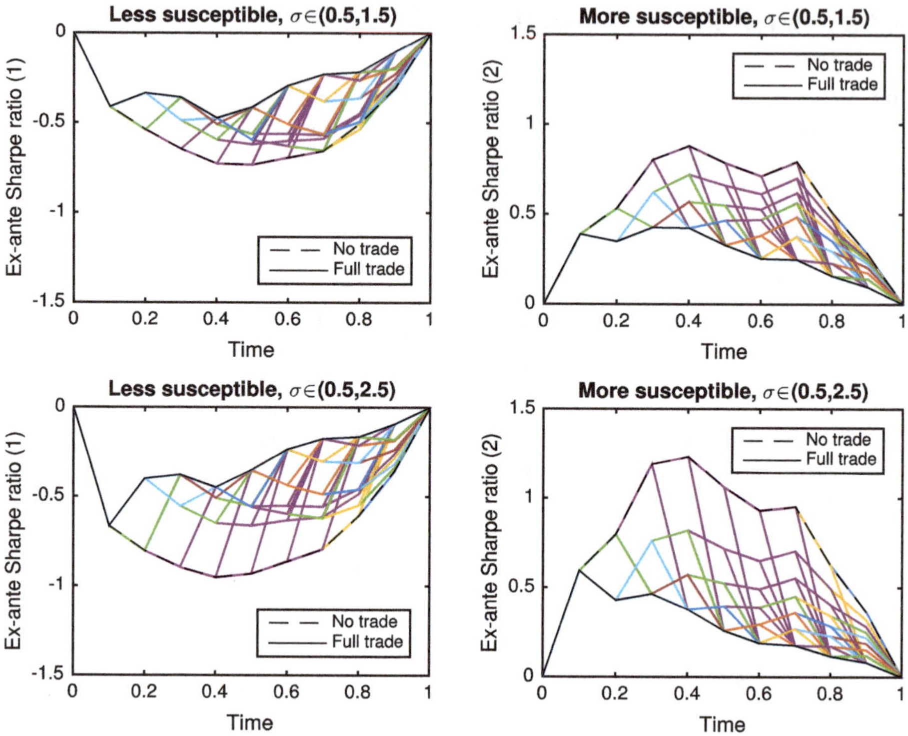

Fig. 3.11 Evolution of Sharpe ratio based on $\mathbb{V}_t^j(\Pi_t^j)$ given in Eq. (3.56) for sample trajectories of ξ_t^1 and ξ_t^2 and all possible trading strategies where dividend is Gaussian

that the following two cases are equivalent: (a) terminal payoff is exogenous (as in our case) and agents are risk-averse, (b) dividend is endogenous and agents are risk-neutral. When the asset dividend (or terminal payoff) is exogenous and agents' actions have no impact on it, one needs to introduce either trade quotas or proper risk aversion assumptions to prevent agents from trading unlimited amounts to make infinite profits, should quoted prices be in their favour. In the presence of informational differences, there would be less or no motivation for agents who are not only informationally less capable but also risk averse to actively participate in a market where the participants are assumed to be rational. Such state of affairs can, in fact, exacerbate the situations where markets shut down due to perceived differential information. Such situations are avoided in the literature by introducing the concept of 'noise-traders' (refer to Table 3.1), which we avoid in the present context so as to focus solely on the influence of differential information on market phenomena.

We assume that agents are risk-averse with the utility assigned to a sure dividend x, i.e.,

$$U_j(x) = -e^{-\lambda_j x} \quad (\lambda_j > 0), \tag{3.57}$$

that is characterised by a constant absolute risk aversion level λ_j. We note that the utility function $U : (0, \infty) \rightarrow \mathbb{R}$ in Eq. (3.57) is C^2, and satisfies $U' > 0$, $U'' < 0$ as well as the Inada conditions [17]. Under U, the certainty equivalent of $\mathbb{E}[X]$ for agent j is given by

$$x_c^j = -\frac{\ln\left(-\mathbb{E}\left[U_j(X)|\bar{\sigma}(\xi_t^j)\right]\right)}{\lambda_j} \tag{3.58}$$

with $x_c^j < \mathbb{E}[X|\bar{\sigma}(\xi_t^2)]$ following from strict concavity. Assuming again X is normal with $\mathcal{N}(0,1)$, the equilibrium strategy in a market where agents maximise their expected utility from terminal wealth[12] is now associated to the objective function which is analogous to Eqs. (3.48) and (3.55) and given by

$$\arg\max_{\left(q_t^j\right)} \sum_{t=0}^{T} \mathbb{E}\left[U_j(\Pi_t^j)|\bar{\sigma}(\xi_t^j)\right] \tag{3.59}$$

where each signal-based price S_t^j is worked out, this time, according to certainty equivalence relation in Eq. (3.58) as follows (assuming $s_t = 0$ for simplicity):

$$\begin{aligned}
U_j(S_t^j) = \mathbb{E}\left[U_j(\pm X)|\xi_t^j\right] &= -\frac{\int_{\mathbb{X}} e^{-\lambda_j \pm x} e^{-\frac{x^2}{2}} e^{\kappa_t\left(\sigma_j \xi_t^j x - \frac{1}{2}\sigma_j^2 x^2 t\right)} \mathrm{d}x}{\int_{\mathbb{X}} e^{-\frac{x^2}{2}} e^{\kappa_t\left(\sigma_j \xi_t^2 x - \frac{1}{2}\sigma_j^2 x^2 r\right)} \mathrm{d}x} \\
&= -\frac{\int_{\mathbb{X}} e^{-\frac{x^2}{2}} e^{\left(\sigma_j \kappa_t \xi_t^j \mp \lambda_j\right)x - \left(\frac{1}{2}\sigma_j^2 \kappa_t t\right)x^2} \mathrm{d}x}{\int_{\mathbb{X}} e^{-\frac{x^2}{2}} e^{\left(\sigma_j \kappa_t \xi_t^j\right)x - \left(\frac{1}{2}\sigma_j^2 \kappa_t t\right)x^2} \mathrm{d}x} \\
&= -e^{\frac{(\sigma_j \kappa_t \xi_t^j \mp \lambda_j)^2}{2(\sigma_j^2 \kappa_t t + 1)}} e^{-\frac{(\sigma_j \kappa_t \xi_t^j)^2}{2(\sigma_j^2 \kappa_t t + 1)}} \\
&= -e^{\frac{1}{2}\frac{\lambda_j^2}{\sigma_j^2 \kappa_t t + 1} \mp \frac{\sigma_j \kappa_t \xi_t^j \lambda_j}{\sigma_j^2 \kappa_t t + 1}} .
\end{aligned} \tag{3.60}$$

This implies, for a bid quote,

$$S_t^j = U_j^{-1}(\mathbb{E}\left[U_j(X)|\xi_t^j\right]) = \frac{\sigma_j \kappa_t \xi_t^j}{\sigma_j^2 \kappa_t t + 1} - \frac{1}{2}\frac{\lambda_j}{\sigma_j^2 \kappa_t t + 1} \tag{3.61}$$

[12] There is no intertemporal consumption.

and, similarly, for an ask quote,

$$S_t^j = -U_j^{-1}(\mathbb{E}\left[U_j(-X)\middle|\xi_t^j\right]) = \frac{\sigma_j \kappa_t \xi_t^j}{\sigma_j^2 \kappa_t t + 1} + \frac{1}{2}\frac{\lambda_j}{\sigma_j^2 \kappa_t t + 1}. \tag{3.62}$$

The second term in Eqs. (3.61) and (3.62) can be considered as the "information-adjusted risk premium" and appears naturally as the bid/ask spread which is inversely proportional to σ_j and t. Thus, given λ_j and t, the more an agent is informationally more (less) susceptible, the lower (higher) a risk premium he will have.

The central planner, on his side, will set the price transaction price to the one which equalises their individual signal-based expected utility from the transaction, i.e.,

$$\mathbb{E}\left[U\left(q_t^1(X - S_t^*)\right)|\xi_t^1\right] = \mathbb{E}\left[U\left(q_t^2(X - S_t^*)\right)|\xi_t^2\right]. \tag{3.63}$$

Assuming again $|q_t^j| \in \{0, 1\}$, market is a high-type and agent 2 buys, the pricing rule in Eq. (3.63) can be arranged further as follows:

$$\begin{aligned}
-\frac{\int_{\mathbb{X}} e^{-\lambda_1(S_t^*-x)} e^{-\frac{x^2}{2}} e^{\kappa_t\left(\sigma_1\xi_t^1 x - \frac{1}{2}\sigma_1^2 x^2 t\right)} \mathrm{d}x}{\int_{\mathbb{X}} e^{-\frac{x^2}{2}} e^{\kappa_t\left(\sigma_1\xi_t^1 x - \frac{1}{2}\sigma_1^2 x^2 t\right)} \mathrm{d}x} &= -\frac{\int_{\mathbb{X}} e^{-\lambda_2(x-S_t^*)} e^{-\frac{x^2}{2}} e^{\kappa_t\left(\sigma_2\xi_t^2 x - \frac{1}{2}\sigma_2^2 x^2 t\right)} \mathrm{d}x}{\int_{\mathbb{X}} e^{-\frac{x^2}{2}} e^{\kappa_t\left(\sigma_2\xi_t^2 x - \frac{1}{2}\sigma_2^2 x^2 r\right)} \mathrm{d}x} \\
-\frac{e^{-\lambda_1 S_t^*} \int_{\mathbb{X}} e^{-\frac{x^2}{2}} e^{(\sigma_1\kappa_t\xi_t^1 + \lambda_1)x - \left(\frac{1}{2}\sigma_1^2\kappa_t t\right)x^2} \mathrm{d}x}{\int_{\mathbb{X}} e^{-\frac{x^2}{2}} e^{(\sigma_1\kappa_t\xi_t^1)x - \left(\frac{1}{2}\sigma_1 j^2\kappa_t t\right)x^2} \mathrm{d}x} &= -\frac{e^{\lambda_2 S_t^*} \int_{\mathbb{X}} e^{-\frac{x^2}{2}} e^{(\sigma_2\kappa_t\xi_t^2 - \lambda_2)x - \left(\frac{1}{2}\sigma_2^2\kappa_t t\right)x^2} \mathrm{d}x}{\int_{\mathbb{X}} e^{-\frac{x^2}{2}} e^{(\sigma_2\kappa_t\xi_t^2)x - \left(\frac{1}{2}\sigma_2^2\kappa_t t\right)x^2} \mathrm{d}x} \\
-e^{-\lambda_1 S_t^*} e^{\frac{(\sigma_1\kappa_t\xi_t^1 + \lambda_1)^2}{2(\sigma_1^2\kappa_t t + 1)}} e^{-\frac{(\sigma_1\kappa_t\xi_t^1)^2}{2(\sigma_1^2\kappa_t t + 1)}} &= -e^{\lambda_2 S_t^*} e^{\frac{(\sigma_2\kappa_t\xi_t^2 - \lambda_2)^2}{2(\sigma_2^2\kappa_t t + 1)}} e^{-\frac{(\sigma_2\kappa_t\xi_t^2)^2}{2(\sigma_2^2\kappa_t t + 1)}} \\
-e^{-\lambda_1 S_t^* + \frac{1}{2}\frac{\lambda_1^2}{\sigma_1^2\kappa_t t + 1} + \frac{\sigma_1\kappa_t\xi_t^1\lambda_1}{\sigma_1^2\kappa_t t + 1}} &= -e^{\lambda_2 S_t^* + \frac{1}{2}\frac{\lambda_2^2}{\sigma_2^2\kappa_t t + 1} - \frac{\sigma_2\kappa_t\xi_t^2\lambda_2}{\sigma_2^2\kappa_t t + 1}},
\end{aligned} \tag{3.64}$$

which directly implies, also by Eq. (3.61), that

$$\begin{aligned}
S_t^* &= \frac{\lambda_1\left(\frac{\sigma_1\kappa_t\xi_t^1}{\sigma_1^2\kappa_t t + 1} + \frac{1}{2}\frac{\lambda_1}{\sigma_1^2\kappa_t t + 1}\right) + \lambda_2\left(\frac{\sigma_2\kappa_t\xi_t^2}{\sigma_2^2\kappa_t t + 1} - \frac{1}{2}\frac{\lambda_2}{\sigma_2^2\kappa_t t + 1}\right)}{\lambda_1 + \lambda_2} \\
&= \frac{\lambda_1}{\lambda_1 + \lambda_2} S_t^1 + \frac{\lambda_2}{\lambda_1 + \lambda_2} S_t^2.
\end{aligned} \tag{3.65}$$

Thus, the information-based market price is the weighted average of risk-adjusted signal-based prices with respect to the risk aversion levels λ_j. Accordingly, a_t^s and b_t^s, and all relevant quantities such as $P_t^j(\xi_c^j|x+,s)$ will need to be updated to re-explore an optimal strategy.

References

1. Admati A, Pfleiderer P (1988) A theory of intraday patterns: volume and price variability. Rev Financ Stud 1(1):3–40
2. Akerlof G (1970) The market for "lemons": quality uncertainty and the market mechanism. Q J Econ 84(3):488–500 .
3. Back K (1992) Insider trading in continuous time. Rev Financ Stud 5(3):387–409
4. Back K, Baruch S (2004) Information in securities markets: Kyle meets glosten and milgrom. Econometrica 72(2):433–465
5. Back K, Pedersen H (1998) Long-lived information and intra-day patterns. J Finan Markets 1:385–402
6. Bond P, Eraslan H (2010) Information-based trade. J Econ Theory 145:1675–1703
7. Brody D, Davis M, Friedman R, Hughston L (2009) Informed traders. Proc R Soc A 465:1103–1122
8. Brody D, Hughston L, Macrina A (2011) Modelling information flows in financial markets. Advanced mathematical methods for finance. Springer, Berlin, pp 133–153
9. Brown A, Rogers L (2012) Diverse beliefs. Stochastics: Int J Probab Stoch Process 84(5–6):683–703
10. Brunnermeier M (2001) Asset pricing under asymmetric information: bubbles, crashes, technical analysis, and herding. Oxford University Press, Oxford
11. Buss A, Dumas B (2013) Financial-market equilibrium with friction. Working Paper 19155, National Bureau of Economic Research (2013)
12. Caldentey R, Stacchetti E (2010) Insider trading with a random deadline. Econometrica 78(1):245–283
13. Cvitanić J, Jouini E, Malamud S, Napp C (2012) Financial markets equilibrium with heterogeneous agents. Rev Financ 16(1):285–321
14. Dumas B, Lyasoff A (2012) Incomplete-market equilibria solved recursively on an event tree. J Financ 67(5):1897–1941
15. Glosten L, Milgrom P (1985) Bid, ask and transaction prices in specialist market with heterogeneously informed traders. J Financ Econ 14:71–100
16. Grossman S, Stiglitz J (1980) On the impossibility of informationally efficient markets. Am Econ Rev 70(3):393–408
17. Inada K-I (1963) On a two-sector model of economic growth: comments and a generalization. Rev Econ Stud 30(2):119–127
18. Kyle A (1985) Continuous auctions and insider trading. Econometrica 53(6):1315–1335
19. Milgrom P, Stokey N (1982) Information, trade and common knowledge. J Econ Theory 26(1):17–27
20. Sharpe W (1966) Mutual fund performance. J Bus 39(1, Part 2: Supplement on Security Prices):119–138

Chapter 4
Putting Signal-Based Model to Work

Building upon the concepts introduced in the previous chapters, this chapter aims to focus on a more practical issue: how can we put signal-based framework to work for a certain type of risky financial asset when there is a multiplicity of real-time market signals on X which, in turn, determines $\phi(X)$. In other words, this chapter illustrates, in a practical sense, how the problem of predicting the future value of a stochastic random variable can be simplified to interpreting the information concerning one of its constituents. In particular, we focus on equity market due to the relatively simpler interpretation of the term "fundamental," referring to the value of a cash-generating business.

No matter how an information flow pattern is modelled, the aim of financial modelling based on forward-looking information is to ensure that future information (e.g., earnings and/or dividend payments in the case of equity) on an asset's fundamental value is represent—either fully or partial—in price discovery.

4.1 Multiple Dividends: Single Market Factor

The risky asset is now characterised at any time t by an infinite number of cashflows which accrue continuously but are announced (or physically distributed) at discrete times intervals. We then call a time-varying subset of these cashflows, i.e., $\{\phi_k\}_{k=1,\ldots,n_t}$, which are due $T_1, \ldots, T_{n_t}$, the cashflows "within" the horizon. Each payoff ϕ_k can be deemed a function of m_k market factors as a subset of $\{X_1, \ldots, X_{\max(m_k)}\}$ for any k, thereby making the price a function of $\max(m_k)$ market factors. This setting indeed allows one to consider a broader spectrum of financial instruments. When each market factor X is associated with an information process $\{\xi_t\}_{0\leq t\leq T_k}$, the problem of valuing an equity reduces to identifying a set of potential candidates for X and, therefore, ξ in a real-world setting, and calibrating the signal flow rate.

N.S. Aydın, *Financial Modelling with Forward-looking Information*, Contributions to Management Science, DOI 10.1007/978-3-319-57147-8_4

For simplicity, we shall assume $m_k = 1$, $\forall k$, throughout the chapter (i.e, a single market factor X determines each cashflow ϕ_k). We further assume that $X_1, \ldots, X_{n_t}$, are i.i.d. At any time t, the σ-algebra $\mathcal{F}_t^{\xi}$ is assumed to be the 'join' of the σ-algebras generated by n_t independent information processes, i.e., $\sigma(\xi_1) \vee \cdots \vee \sigma(\xi_{n_t})$, and the current and past values of the market factors and the risky asset, i.e., $\mathcal{F}_t = \sigma(X_s, S_s : 0 \leq s \leq t)$:

$$\mathcal{F}_t^{\xi} := \sigma(\xi_1) \vee \cdots \vee \sigma(\xi_{n_t}) \vee \mathcal{F}_t. \tag{4.1}$$

The price of the asset is then simply given by

$$S_t = \sum_{k=1}^{n_t} \mathbf{1}_{\{t<T_k\}} e^{-r(T_k-t)} \mathbb{E}\left[\phi_{T_k}(X) | \mathcal{F}_t^{\xi}\right], \tag{4.2}$$

where $\mathbb{E}$ is w.r.t. $\mathbb{Q}$ by setting. Furthermore, the dynamics of asset price process $\{S_t\}_{t\geq 0}$ is analogous to those derived in Chap. 2 for the single cashflow case (cf. [4]):

$$\begin{aligned} \mathrm{d}S_t &= r_t S_t \mathrm{d}t \\ &\quad + \mathbf{1}_{\{t<T_1\}} e^{-r(T_1-t)} \sigma_1 \kappa_t^1 \mathbb{C}\mathrm{ov}_t \left[\phi(X_{T_1}), X_{T_1}\right] \mathrm{d}W_t^1 \\ &\quad \vdots \\ &\quad + \mathbf{1}_{\{t<T_n\}} e^{-r(T_n-t)} \sigma_n \kappa_t^n \mathbb{C}\mathrm{ov}_t \left[\phi(X_{T_n}), X_{T_n}\right] \mathrm{d}W_t^n \\ &= r_t S_t \mathrm{d}t + \sum_{k=1}^{n_t} \mathbf{1}_{\{t<T_1\}} e^{-r(T_k-t)} \sigma_k \kappa_t^k \mathbb{C}\mathrm{ov}_t \left[\phi(X_{T_k}), X_{T_k}\right] \mathrm{d}W_t^k \\ &\quad - \sum_{k=1}^{n_t} \phi(X_{T_k}) \mathrm{d}\mathbf{1}_{\{t>T_k\}}, \end{aligned} \tag{4.3}$$

where $\mathbb{C}\mathrm{ov}$ is the covariance function. The last term in Eq. (4.3) comes from the price adjustment due to accrual of cashflow (ex-dividend). Equation (4.3) implies that the asset price dynamics in the multiple cashflow case based on signal-based framework remains fairly tractable. In what follows, we show how the present framework can be applied on real market data with slight modifications.

4.2 The Case for "Implied" Dividends

In [5] and [23], the present concept is applied to produce a tractable formula for storable commodity prices under the assumption that the asset pays—what authors call—a continuous 'convenience dividend' that is assumed to follow

Ornstein–Uhlenbeck (OU) dynamics. The OU process is then associated, through its orthogonal decomposition, to the concept of 'OU bridge,' thereby putting together analytical formulae for commodity spot and derivatives prices. In this chapter, we introduce the concept of "implied dividend," which is based on earnings, as the stochastic market factor X that determines the dividend through identity $\phi(X) = X$ and, eventually, the equity price.

There is indeed a large body of literature which argues that it is reasonable to consider earnings data as a proxy for company's expected dividends and measures tied to the former, rather than the latter, will likely provide better information about the actual cash flows generated (see, e.g., [6, 19, 20]). Indeed, many growth businesses choose not to pay cash dividends but, instead, to use their earnings to repurchase outstanding shares or to reinvest in future expansion—making earnings a more informative measure of the fundamental value of a business (see [6, 7]). Investors are also far more interested in the earnings potential of a business rather than its paid dividends (cf. [2]). Earnings, like dividends, are also generated on a continuous basis, although their true value is revealed at discrete time points (quarterly or annually), justifying their suitability for use in continuous-time setups.

In the sequel, we assume that earnings are the basis for changes in an asset's value as "invisible" dividends and they provide some kind of convenience yield $\phi(X)$ which become known to agents at T_k, $k = 1, \ldots, n$. The raw signal process ξ_t in this case conveys noisy information about the true value of the earnings (and, eventually, dividends).

Based on the Bakshi–Chen model introduced in [2], we relate earnings X to "implied" dividends X' as follows:

$$\phi(X'_k) = \delta\phi(X_k) + \epsilon_k, \quad \epsilon_k \sim N(0, \sigma_k^{\epsilon}). \tag{4.4}$$

where we implicitly assumed ϕ to be identity. Here, $\delta \in [0, 1]$ is the dividend payout ratio. The use of constant payout ratio is common in the equity valuation literature. The classic survey in [18] finds that indeed $\delta_t \to \delta$. The rationale for and interpretation behind Eq. (4.4) is addressed in [2] and [7]. Therefore, independent of whether the firm pays cash dividends or not, we interpret $\delta\phi(X)$ given in Eq. (4.4) as the "implied dividend" which will be the governing factor X in our model behind asset price movements.

Furthermore, we assume that earnings and earnings growth follow a geometric Brownian and Ornstein-Uhlenbeck (OU) dynamics, respectively. That is,

$$\mathrm{d}X_k = \mu_k^X X_k \mathrm{d}T_k + \sigma_X \mathrm{d}W_k^X, \tag{4.5}$$

$$\mathrm{d}\mu_k^X = \alpha(\mu_0 - \mu_k^X)\mathrm{d}T_k + \sigma_\mu \mathrm{d}W_k^{\mu}, \tag{4.6}$$

where $\mathrm{d}T_k = T_k - T_{k-1}$. Also we set $W_k^X \perp\!\!\!\perp W_k^{\mu}$ for $\mathcal{F}^{\xi}$-adapted W_k^X and W_k^{μ} that are martingale under the pricing measure. In what follows, we will be employing this model to estimate the parameters of our signal-based valuation framework.

4.2.1 *Recovering the Gordon Model in Continuous Time*

First we associate the signal-based price S_t to Gordon model [10, 11] under constant earnings growth assumption, and later on extend this to time-varying growth.

4.2.1.1 Constant Earnings Growth

Assuming S pays an infinite strip of earnings starting from u, where $u > t$, the continuous time analogous of Eq. (4.2) is

$$S_t = e^{-r_t^u(u-t)} \int_u^{\infty} e^{-r_b(v-u)} \mathbb{E}\left[\delta X_v | \mathcal{F}_t^{\xi}\right] \mathrm{d}v, \tag{4.7}$$

where $\mathcal{F}_t^{\xi}$, $\mathcal{F}_t$ are as given in Eq. (4.1). When μ^X is constant, say μ_0, a straightforward calculation yields

$$\begin{aligned}
S_t &= \delta e^{-r_t^u(u-t)} \int_u^{\infty} e^{-r_b(v-u)} \mathbb{E}\left[X_u e^{(\mu_0 - \frac{1}{2}\sigma_X^2)(v-u)+\sigma_X W_{v-u}^X} \Big| \mathcal{F}_t^{\xi}\right] \mathrm{d}v \\
&= \delta e^{-r_t^u(u-t)} \phi_t(X_u) \int_u^{\infty} e^{-r_b(v-u)} \mathbb{E}\left[e^{(\mu_0 - \frac{1}{2}\sigma_X^2)(v-u)+\sigma_X W_{v-u}^X}\right] \mathrm{d}v \\
&= \delta e^{-r_t^u(u-t)} \phi_t(X_u) \int_u^{\infty} e^{-r_b(v-u)} e^{(\mu_0 - \frac{1}{2}\sigma_X^2)(v-u)+\frac{1}{2}\sigma_X^2(v-u)} \mathrm{d}v \\
&= \delta e^{-r_t^u(u-t)} \phi_t(X_u) \int_u^{\infty} e^{-(r_b-\mu_0)(v-u)} \mathrm{d}v \qquad (\eta = v-u) \\
&= \delta e^{-r_t^u(u-t)} \phi_t(X_u) \int_0^{\infty} e^{-(r_b-\mu_0)\eta} \mathrm{d}\eta \\
&= \delta e^{-r_t^u(u-t)} \frac{\phi_t(X_u)}{r_b-\mu_0} \qquad (r_b > \mu_0),
\end{aligned} \tag{4.8}$$

where r_b is the investment benchmark while r_t^u is the money market rate for maturity u. Equation (4.8) is nothing but the earnings (or implied dividend) equivalent of the well-known intrinsic value model of Gordon. Note the slight difference in appearance between the discrete and continuous forms of the Gordon model (cf. [16]) which disappears as $(1 + \mu_0 \mathrm{d}\eta)\phi_t(X_u) \to \phi_t(X_u)$ as $\mathrm{d}\eta \to 0$. This model, however, is mostly criticised for assuming that the dividend growth rate as well as the risk-adjusted discount rate remain constant—a point which is confronted in the literature by the well-known St. Petersburg paradox (see, e.g., [8]). In our pricing algorithm, we shall circumvent this issue by considering a constant spread between μ_0 and r_b.

4.2.1.2 Time-varying Earnings Growth

If μ^X were time-variant, on the other hand, we would simply have an additional term $\exp(\int_t^u \mu_\nu \mathrm{d}\nu)$ substituting for μ_0 from the first line of Eq. (4.8). By the well-known solution to the OU process in Eq. (4.6), we have

$$\mu_\nu \sim \mathcal{N}\left(\mu_t e^{-\alpha(\nu-t)} + \mu_0(1-e^{-\alpha(\nu-t)}), \left(\frac{\sigma_\mu^2}{2\alpha}(1-e^{-2\alpha(\nu-t)})\right)^{1/2}\right), \tag{4.9}$$

with $t \leq \nu \leq u$. This implies

$$\begin{aligned}\mathbb{E}\left[\exp\left(\int_t^u \mu_\nu \mathrm{d}\nu\right)\right] = \exp\left(\int_t^u \mu_t e^{-\alpha(\nu-t)} + \mu_0(1-e^{-\alpha(\nu-t)})\mathrm{d}\nu\right)\cdot \\ \mathbb{E}\left[\exp\left(\int_t^u \mu'_\nu \mathrm{d}\nu\right)\right],\end{aligned} \tag{4.10}$$

where the notation μ'_ν is introduced to denote μ_ν without a drift. The variance of $\int_t^u \mu'_\nu \mathrm{d}\nu$ is given by

$$\mathbb{V}\left(\int_t^u \mu'_\nu \mathrm{d}\nu\right) = \mathbb{E}\left[\int_t^u \mu'_\nu \mathrm{d}\nu_1 \int_t^u \mu'_\nu \mathrm{d}\nu_2\right] = \int_t^u \int_t^u \mathbb{E}\left[\mu'_{\nu_1}\mu'_{\nu_2}\right] \mathrm{d}\nu_1 \mathrm{d}\nu_2. \tag{4.11}$$

The covariance term $\mathbb{E}\left[\mu'_{\nu_1}\mu'_{\nu_2}\right]$ follows from the solution to the driftless OU process, i.e.,

$$\begin{aligned}\mathbb{E}\left[\mu'_{\nu_1}\mu'_{\nu_2}\right] &= \mathbb{E}\left[\int_t^{\nu_1} e^{-\alpha(\nu_1-s_1)}\mathrm{d}W_{s_1}^{\mu_1} \int_t^{\nu_2} e^{-\alpha(\nu_2-s_2)}\mathrm{d}W_{s_2}^{\mu_2}\right] \\ &= \sigma_\mu^2 e^{-\alpha(\nu_1+\nu_2)}\mathbb{E}\left[\int_t^{\nu_1} e^{\alpha s_1}\mathrm{d}W_{s_1}^{\mu_1} \int_t^{\nu_2} e^{\alpha s_2}\mathrm{d}W_{s_2}^{\mu_2}\right] \\ &= \sigma_\mu^2 e^{-\alpha(\nu_1+\nu_2)}\mathbb{E}\left[\int_0^{\nu_1-t} e^{\alpha(u_1+t)}\mathrm{d}W_{u_1+t}^{\mu_1} \int_0^{\nu_2-t} e^{\alpha(u_2+t)}\mathrm{d}W_{u_2+t}^{\mu_2}\right] \\ &= \sigma_\mu^2 e^{-\alpha(\nu_1+\nu_2-2t)}\mathbb{E}\left[\int_0^{\nu_1-t} e^{\alpha u_1}\mathrm{d}W_{u_1}^{\mu_1} \int_0^{\nu_2-t} e^{\alpha u_2}\mathrm{d}W_{u_2}^{\mu_2}\right] \\ &= \frac{\sigma_\mu^2}{2\alpha} e^{-\alpha(\nu_1+\nu_2-2t)}\left(e^{2\alpha\min(\nu_1-t,\nu_2-t)} - 1\right) \\ &= \frac{\sigma_\mu^2}{2\alpha} e^{-\alpha(\nu_1+\nu_2)}\left(e^{2\alpha\min(\nu_1,\nu_2)} - 1\right).\end{aligned} \tag{4.12}$$

Simply by assuming $\nu_1 < \nu_2$, w.l.o.g., $\mathbb{V}\left(\int_t^u \mu'_\nu \mathrm{d}\nu\right)$ can be found as

$$\begin{aligned}
\mathbb{V}\left(\int_t^u \mu'_\nu \mathrm{d}\nu\right) &= \int_t^u \int_t^u \frac{\sigma_\mu^2}{2\alpha}\left(e^{-\alpha(\nu_2-\nu_1)} - e^{-\alpha(\nu_2+\nu_1)}\right)\mathrm{d}\nu_2\mathrm{d}\nu_1 \\
&= \frac{\sigma_\mu^2}{2\alpha}\int_t^u \left(\frac{e^{-\alpha(\nu_2-\nu_1)}}{-\alpha} - \frac{e^{-\alpha(\nu_2+\nu_1)}}{-\alpha}\right)\Bigg|_t^u \mathrm{d}\nu_1 \\
&= \frac{\sigma_\mu^2}{2\alpha^2}\int_t^u \left(e^{-\alpha(u+\nu_1)} - e^{-\alpha(u-\nu_1)}\right) - \left(e^{-\alpha(t+\nu_1)} - e^{-\alpha(t-\nu_1)}\right)\mathrm{d}\nu_1 \\
&= \frac{\sigma_\mu^2}{2\alpha^2}\int_t^u e^{-\alpha u}\left(e^{-\alpha\nu_1} - e^{\alpha\nu_1}\right) - e^{-\alpha t}\left(e^{-\alpha\nu_1} - e^{\alpha\nu_1}\right)\mathrm{d}\nu_1 \\
&= \frac{\sigma_\mu^2}{2\alpha^2}\left(e^{-\alpha u} - e^{-\alpha t}\right)\int_t^u \left(e^{-\alpha\nu_1} - e^{\alpha\nu_1}\right)\mathrm{d}\nu_1 \\
&= \frac{\sigma_\mu^2}{2\alpha^2}\left(e^{-\alpha u} - e^{-\alpha t}\right)\left(\frac{e^{-\alpha\nu_1}}{-\alpha} - \frac{e^{\alpha\nu_1}}{\alpha}\right)\Bigg|_t^u \\
&= \frac{\sigma_\mu^2}{2\alpha^3}\left(e^{-\alpha u} - e^{-\alpha t}\right)\left[\left(e^{-\alpha u} - e^{\alpha u}\right) - \left(e^{-\alpha t} - e^{\alpha t}\right)\right] \\
&= \frac{\sigma_\mu^2}{2\alpha^3}\left[2e^{-\alpha(u+t)} + e^{-\alpha(u-t)} + e^{\alpha(u-t)} - e^{-2\alpha u}e^{-2\alpha t} - 2\right]. \qquad (4.13)
\end{aligned}$$

Returning back to Eq. (4.10), we conclude

$$\begin{aligned}
\mathbb{E}\left[\exp\left(\int_t^u \mu_\nu \mathrm{d}\nu\right)\right] &= \exp\Bigg[\mu_t\frac{1}{\alpha}\left(1 - e^{-\alpha(u-t)}\right) + \mu_0\left((u-t) - \frac{1}{\alpha}\left(1 - e^{-\alpha(u-t)}\right)\right) \\
&\quad + \frac{1}{2}\frac{\sigma_\mu^2}{2\alpha^3}\left(2e^{-\alpha(u+t)} + e^{-\alpha(u-t)} + e^{\alpha(u-t)} - e^{-2\alpha u}e^{-2\alpha t} - 2\right)\Bigg].
\end{aligned} \qquad (4.14)$$

Equation (4.14) would be accommodated into Eq. (4.8) to derive a time-varying growth version of the Gordon model. This, however, is beyond the scope of our analysis in this chapter. Below, we introduce the earnings signals that will act, among possible others, as our information flow process ξ_t. We also introduce a slightly modified version of the latter.

4.3 Real-Time Information Flow

Financial markets, with equity market being a particular example, are forward-looking, i.e., prices are ideally discovered on the basis of expectations pertaining to the future value-generating ability of the underlying business. One vivid example to this is the price adjustments to an equity following unexpected deviations of realised earnings from their consensus values and/or inter-temporal revisions of earnings expectations by brokers. We explain how this property can be worked to fit it into the present context in more detail below.

At time t_k, market experts start disseminating their consensus estimates on the true value of a certain ticker's quarterly earnings value X_k, and therefore its implied dividend, which is due at T_k. These consensus figures are derived from comprehensive assessments of up to 40 brokerage analysts which closely follow a certain ticker and incorporate as much information as available. As for quarterly earnings consensus, $T_k - t_k$ is generally between 2 and 4 years. We exhibit in Fig. 4.1 the quarterly earnings signals extracted from Bloomberg terminal for a large-cap U.S. blue-chip company (ticker: `MSFT`) that are released at an average frequency of four days during 2004Q1–2015Q3. The data are adjusted for corporate issues such as stock splits, exclude non-recurring items, include employee stock options expenses, and incorporate any guidance issued by the company prior to actual earnings announcement.

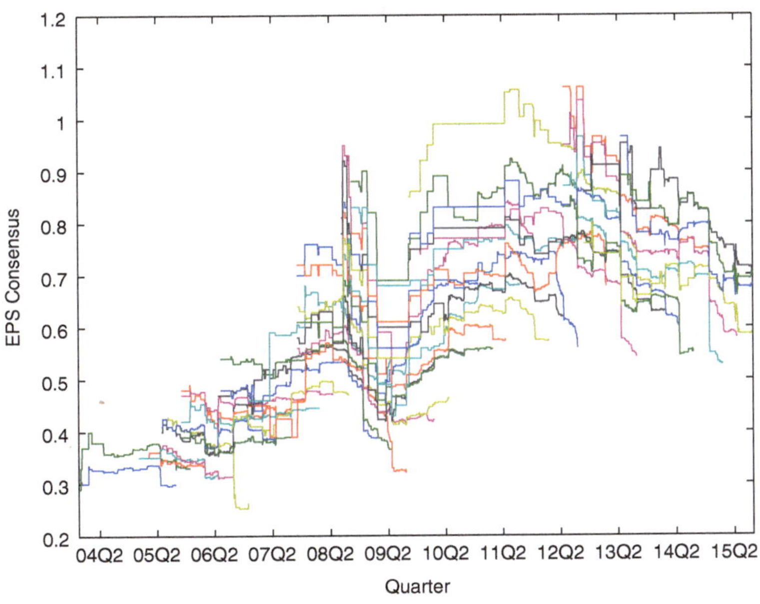

Fig. 4.1 Evolution of quarterly earnings signals. Data source: Bloomberg

As we cannot separate X from noise in any observed signal to construct empirically the desired ξ given in Eq. (2.4),[1] we introduce[2] a slightly modified version of ξ in Eq. (2.4), while preserving its intuitive properties, as follows:

$$\xi_t^k = \begin{cases} x_k^* + \tau_t^k(X_k - x_k^*) + \frac{\beta_t^k}{\sigma(T_k - t_k)}, & \text{if} \quad t_k \le t \le T_k, \\ \emptyset, & \text{otherwise}, \end{cases} \tag{4.15}$$

where x_k^* is the first signal sample received at time t_k about the true value of X_k, σ^{-1} now a measure of noise-to-signal, and

$$\tau_t^k = \mathbf{1}_{t \le t_k} \frac{t - t_k}{T_k - t_k} \in [0, 1] \tag{4.16}$$

the proportion of signal lifetime elapsed since it started being transmitted. Intuitively, we now allow an increasing σ to suppress noise (thereby, increasing signal-to-noise) rather than to increase the signal content directly, as in Eq. (2.4). The conditional variance of ξ_t^k given $X_k = x$ can be rewritten as

$$\mathbb{V}\left(\xi_t^k | X_k = x\right) = \frac{1}{\sigma^2} \frac{(t - t_k)(T_k - t)}{(T_k - t_k)^3} = \frac{\tau_t^k\left(1 - \tau_t^k\right)}{\sigma^2(T_k - t_k)}. \tag{4.17}$$

Note that the modified version of the information process ξ_t in Eq. (4.15) is better suited to the real-time signals considered here and, again, do not compromise intuitive and statistical properties of ξ_t described in Chap. 2. The equivalence of the latter two in the sense of integral in Eq. (2.19) can easily be seen as follows. Let us assume w.l.o.g. that $x^* = 0$ and $t_k = 0$. Then, apparently,

$$\int_A xp(x)e^{-\frac{1}{2}\left(\frac{\xi - ax}{b}\right)^2} \mathrm{d}x = \int_A xp(x)e^{-\frac{1}{2}\left(\frac{\xi' - ax/c}{b/c}\right)^2} \mathrm{d}x \tag{4.18}$$

with $a = \sigma t$, $b = \tau_t(T - t)$, $\xi' = \xi/c$, and $c = \sigma T$.

As indicated by Eq. (4.15), the σ-algebra F_t^ξ constantly enlarges and shrinks whenever the number of available signals increases and decreases, respectively. Once the signal ξ^k is started to be received at time t_k, the market updates its prior information about X_k (i.e., $p_X(x)$) through relation (2.14). On the other hand, the noise-to-signal measure $1/\sigma$ needs to be determined from the data. Again, for $t > T_k$, i.e., once X_k has been revealed, $(\xi_t)_{t_1 \le t \le T}$ becomes degenerate (information-null).

---

[1] For instance, one does not normally observe a noisy signal for $\sigma t X$ in the market but X and, therefore, the desired signal $\sigma t X + \epsilon_t$ cannot be recovered from $X + \epsilon_t$.

[2] Also based on recommendation from Edward Hoyle, PhD, Department of Mathematics, Imperial College London.

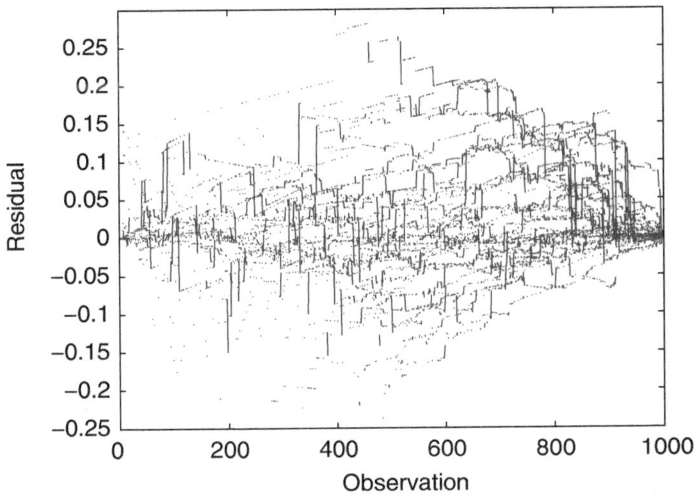

Fig. 4.2 Residuals of empirical information processes depicted in Fig. 4.1

To check the boundary values, apparently, $\xi^k_{t_k} = x^*_k$ and $\xi^k_{T_k} = X_k$, with the latter ensuring that the marginal law of ξ^k_t is the a priori law of X_k (cf. [13, 14]). In Fig. 4.2, we plot the residuals β^k_t from several paths of actual earnings signals, extracted as per Eq. (4.15) whereas their starting and end points are aligned.[3] Sample residuals do indeed exhibit properties that are similar to those of a bridge process. Furthermore, jumps occur occasionally as a result of the significant revisions of consensus data.

4.4 Calibrating the Information Flow Rate

The information flow parameter σ_k, which is time-homogeneous by our setting, is calibrated based on the modified information process given in Eq. (4.15) as follows. We have an sample history of $N = 41$ quarterly earnings signals (with lengths varying from 1.1 to 4.7 years) for the stock ticker considered.[4] To calibrate σ_k,

1. We first extract the linear part of each signal according to Eq. (4.15) to get various paths of the empirical bridge processes. We refer the reader back to Fig. 4.2 for a visualisation of the residual series β^k, $k = 1, \ldots, N$.

[3]We recall that $T_k - t_k$ differs across signals.

[4]The first signal in sample commences on February 20, 2004, and lasts until on October 27, 2005, pinned to 2005Q3 earnings per share figure, whereas the last signal starts and ends on November 29, 2012, and October 21, 2015, respectively, pinned to 2015Q3 earnings per share figure.

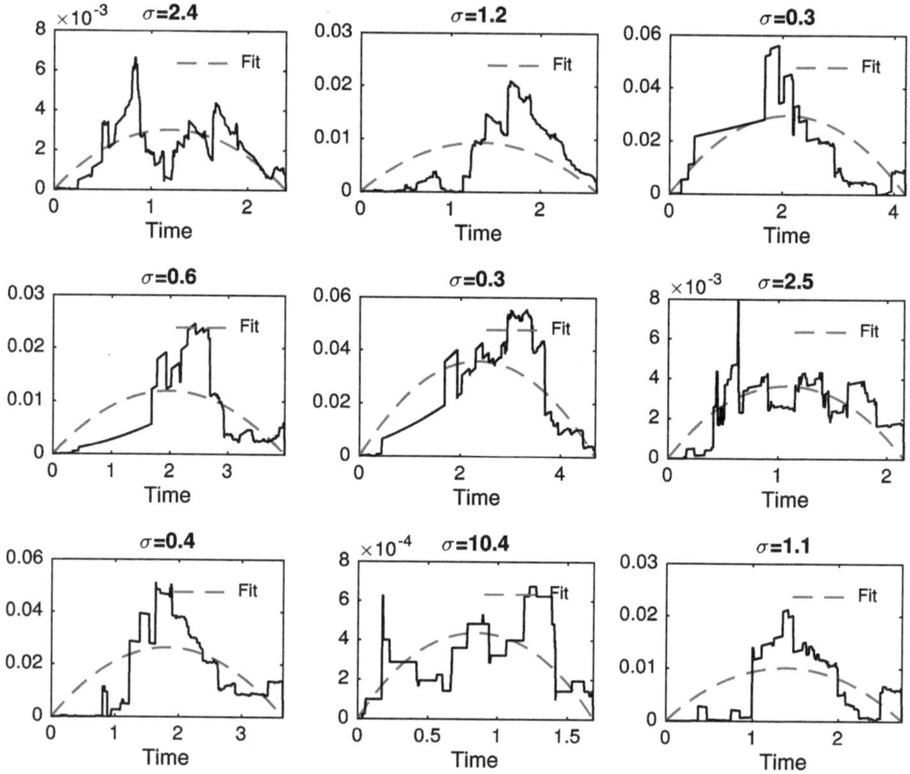

Fig. 4.3 Calibration results of information flow rate, i.e., $\hat{\sigma}_k$, using Levenberg–Marquardt nonlinear curve-fitting algorithm. Arbitrary signals are shown

2. For each β^k, we then run the following non-linear regression based on the theoretical variance of β_t^k:

$$\left(\beta_t^k\right)^2 = \frac{1}{\hat{\sigma}_k^2} \frac{\tau_t^k(1-\tau_t^k)}{T_k - t_k} + \epsilon_t^k, \quad t_k \leq t \leq T_k, \tag{4.19}$$

where, again, $\tau_t^k = (t - t_k)/(T_k - t_k)$ and, presumably, $\epsilon_t^k | \mathcal{F}_t^{\xi} \sim \mathcal{N}(0, \sigma_\epsilon)$. Figure 4.3 shows the calibration results of $\hat{\sigma}_k$ for arbitrary quarterly earnings signals, whereby the Levenberg–Marquardt nonlinear curve-fitting algorithm is used.[5,6]

[5] Parameter estimates for all signals are statistically significant with considerably low p-values.

[6] We also remark that some other statistical learning algorithms, e.g., Expectation-Maximisation algorithm (cf. [12]), could also be used to capture possible multi-modal dynamics.

3. (Optional) As a final step, assuming $\beta^1 \perp\!\!\!\perp \ldots \perp\!\!\!\perp \beta^k$, we perform a simple variance averaging over all fitted curves resulting from Eq. (4.19) to find $\hat{\sigma}$:

$$\hat{\sigma} = \left(\frac{\sum_{k=1}^{N} \hat{\sigma}_k^{-2}}{N} \right)^{-1/2}. \tag{4.20}$$

The last step yields $\hat{\sigma} = 0.79$ for the ticker `MSFT` with respect to the period 2005Q3–2015Q3. In what follows, we develop a closed-form approximation to signal-based price and present the pricing results.

4.5 Analytical Approximation to Signal-Based Price

We now turn our attention back to deriving a preferably crisp formula for pricing the risky asset when there is a multiplicity of information processes $\xi_{t_1 T_1}, \ldots, \xi_{t_n T_n}$, delivering a continuum of market signals on i.i.d. market factors $X_1, \ldots, X_n$, and, thereby, cashflows $\phi(X_1), \ldots, \phi(X_n)$, through the identity $\phi(x) = x$.

For simplicity of exposition and without loss of generality, consider any three signals with

$$(T_{k-1}, T_k, T_{k+1}) = (r, s, u), \quad r < s < u. \tag{4.21}$$

We state the well-known solution to the SDE of X in Eq. (4.5) at time u:

$$\begin{aligned} X_u &= X_s \exp\left(\left[\mu_{su}^X - \sigma_X^2/2 \right] (u-s) + \sigma_X W_{u-s}^X \right) \\ &= X_s \exp\left(\left[\mu_{rs}^X e^{-\alpha(s-r)} + \mu_0 (1 - e^{-\alpha(s-r)}) - \sigma_X^2/2 \right] (u-s) \right. \\ &\quad \left. + \sigma_X W_{u-s}^X + \sigma_\mu (u-s) \int_r^s e^{-\alpha(s-v)} \mathrm{d}W_v^\mu \right). \end{aligned} \tag{4.22}$$

where μ_{su} is simply the growth between s to u. Therefore, by virtue of Eq. (4.12) and using the fact that $W^X \perp\!\!\!\perp W^\mu$, we can write X_u/X_s in the conditionally log-normal form

$$Y_{su} = \frac{X_u}{X_s} = \exp\left(\tilde{\mu}_{su}(u-s) + \tilde{\sigma}_{su} Z_{u-s} \right), \tag{4.23}$$

where $Z \sim \mathcal{N}(0, u-s)$, and with

$$\tilde{\mu}_{su} = \mu_{rs}^X e^{-\alpha(s-r)} + \mu_0 (1 - e^{-\alpha(s-r)}) - \sigma_X^2/2 \tag{4.24}$$

and

$$\tilde{\sigma}_{su}^2 = \sigma_X^2 + \frac{\sigma_\mu^2}{2\alpha} (u-s)(1 - e^{-2\alpha(s-r)}). \tag{4.25}$$

The pricing relation at time t, $s \leq t \leq u$, will then be based on

$$S_t = e^{-r_t^u(u-t)}\mathbb{E}\left[\delta X_u + \epsilon_u | \mathcal{F}_t^{\xi}\right] = e^{-r_t^u(u-t)}\mathbb{E}\left[\delta X_u | \mathcal{F}_t^{\xi}\right], \tag{4.26}$$

where, again, $\mathcal{F}_t^{\xi}$ is defined as in Eq. (4.1). The second equality, in fact, follows from the fact that ξ_t carries information only about X_u (i.e., $\epsilon_u \perp\!\!\!\perp \xi_t$) and the assumption that $\mathbb{E}\left[\epsilon_u | \{X_t\}_{t<u}\right] = 0$.

We now know from Eq. (4.22), (4.25) and (4.24) that, conditionally,

$$X_u | X_s \sim \log \mathcal{N}(\tilde{\mu}'_{su}, \tilde{\sigma}_{su}) \tag{4.27}$$

with $\tilde{\mu}'_{su} := \ln X_s/(u-s) + \tilde{\mu}_{su}$. Then, the pricing relation in Eq. (4.26) implies,

$$\begin{aligned} S_t &= e^{-r_t^u(u-t)}\mathbb{E}\left[\delta X | \xi_t\right] \\ &= \delta e^{-r_t^u(u-t)} \\ &\cdot \frac{\displaystyle\int_{\mathbb{X}} \exp\left(-\frac{1}{2}\frac{(\ln x - \tilde{\mu}'_{su}(u-s))^2}{\tilde{\sigma}^2_{su}(u-s)} - \frac{1}{2}\frac{(\xi_t - [x^* + \tau_t(x - x^*)])^2}{\tau_t(1-\tau_t)/(\sigma^2(u-t))}\right) \mathrm{d}x}{\displaystyle\int_{\mathbb{X}} x^{-1}\exp\left(-\frac{1}{2}\frac{(\ln x - \tilde{\mu}'_{su}(u-s))^2}{\tilde{\sigma}^2_{su}(u-s)} - \frac{1}{2}\frac{(\xi_t - [x^* + \tau_t(x - x^*)])^2}{\tau_t(1-\tau_t)/(\sigma^2(u-t))}\right) \mathrm{d}x}, \end{aligned} \tag{4.28}$$

where $\mathbb{X} = (0, \infty)$. Equation (4.28) is apparently not very handy without recourse to numerical methods. We try to circumvent this issue by using two possible analytical approximations, namely, through gamma and log-gamma distributions.

More specifically, this uses either $X \sim \Gamma(a, b)$ approximately, or $Z \sim \log\Gamma(a, b)$, again, approximately, where $Z = \log X$, and Γ and $\log\Gamma$ are gamma and log-gamma[7] probability laws with densities

$$fx(x|a, b) = \frac{x^{a-1}}{b^a\Gamma(a)}e^{-x/b}, \quad x \in (0, \infty), \tag{4.29}$$

and

$$fz(z|a, b) = \frac{1}{b^a\Gamma(a)}e^{az - e^z/b}, \quad z \in (-\infty, \infty), \tag{4.30}$$

[7]"Log-gamma" in the sense that it is logarithm of a gamma random variable (not that its logarithm is a gamma distribution as in the case of, e.g., lognormal distribution).

respectively.[8] For approximation, we choose to minimise the Kullback–Leibler [17] divergence between the theoretical and approximating density functions, i.e.,

$$\arg\min_{(a,b)} \int_{\mathbb{X}} f_X(\tilde{\mu}'_{su}(u-s), \tilde{\sigma}_{su}\sqrt{u-s}) \cdot \log_2\left(\frac{f_X(\tilde{\mu}'_{su}(u-s), \tilde{\sigma}\sqrt{u-s})}{f'_X(a_{su}, b_{su})}\right) \mathrm{d}x, \tag{4.31}$$

or

$$\arg\min_{(a,b)} \int_{\mathbb{Z}} f_Z(\tilde{\mu}'_{su}(u-s), \tilde{\sigma}_{su}\sqrt{u-s}) \cdot \log_2\left(\frac{f_Z(\tilde{\mu}'_{su}(u-s), \tilde{\sigma}\sqrt{u-s})}{f'_Z(a_{su}, b_{su})}\right) \mathrm{d}z, \tag{4.32}$$

which ensures the expected entropic (or, informational) distance between the latter two is minimised. Approximate analytical solution to problems given in Eqs. (4.31) and (4.32), on the other hand, is given by

$$a_{su} \approx \frac{1}{\tilde{\sigma}^2_{su}(u-s)}, \quad b_{su} \approx \tilde{\sigma}^2_{su}(u-s)\exp\left(\left(\tilde{\mu}'_{su} + \frac{\tilde{\sigma}^2_{su}}{2}\right)(u-s)\right) \tag{4.33}$$

(see Appendix A for a sketch of proof). Figure 4.4 shows the results of gamma approximation to the log-normal density for different parameter values, using both numerical and analytical solutions to the Kullback–Leibler minimisation problem. The approximation works extremely good, particularly for small variances values, and this is why it will work particularly good in our context.

We use this property to replace the log-normal density (normal density) with its gamma (log-gamma) conjugate prior with shape and scale parameters $a_{su} = a(\tilde{\mu}'_{su}, \tilde{\sigma}_{su})$ and $b_{su} = b(\tilde{\mu}'_{su}, \tilde{\sigma}_{su})$, which yields

$$\tilde{S}_t = \delta e^{-r^u_t(u-t)} \frac{\int_{\mathbb{X}} x^{a_{su}} \exp\left(-\frac{x}{b_{su}} - \frac{1}{2}\frac{(\xi_t - [x^* + \tau_t(x-x^*)])^2}{\tau_t(1-\tau_t)/(\sigma^2(u-t))}\right) \mathrm{d}x}{\int_{\mathbb{X}} x^{a_{su}-1} \exp\left(-\frac{x}{b_{su}} - \frac{1}{2}\frac{(\xi_t - [x^* + \tau_t(x-x^*)])^2}{\tau_t(1-\tau_t)/(\sigma^2(u-t))}\right) \mathrm{d}x}, \tag{4.34}$$

[8] We remark that gamma distribution is conjugate prior to log-normal distribution with a known mean.

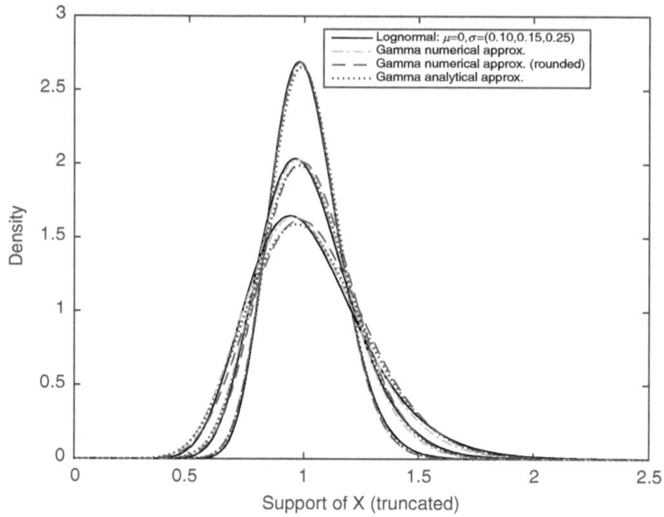

Fig. 4.4 Approximation of conditional log-normal by its conjugate prior gamma

where $\tilde{S}_t$ is the approximation to S_t with a gamma prior and, again, $\mathbb{X} = (0, \infty)$. We can further simplify Eq. (4.34) as follows:

$$\tilde{S}_t = \delta e^{-r_t^u(u-t)} \frac{\int_{\mathbb{X}} x^{a_{su}} \exp\left(-\frac{x}{b_{su}} - \frac{1}{2}\frac{(\xi_t - (1-\tau_t)x^* - \tau x)^2}{\tau_t(1-\tau_t)/(\sigma^2(u-t))}\right) \mathrm{d}x}{\int_{\mathbb{X}} x^{a_{su}-1} \exp\left(-\frac{x}{b_{su}} - \frac{1}{2}\frac{(\xi_t - (1-\tau)x^* - \tau x)^2}{\tau_t(1-\tau_t)/(\sigma^2(u-t))}\right) \mathrm{d}x}$$

$$= \delta e^{-r_t^u(u-t)} \frac{\int_{\mathbb{X}} x^{a_{su}} \exp\left(-\frac{x}{b_{su}} - \frac{1}{2}\frac{-2\tau_t x(\xi_t - (1-\tau_t)x^*) + (\tau x)^2}{\tau_t(1-\tau_t)/(\sigma^2(u-t))}\right) \mathrm{d}x}{\int_{\mathbb{X}} x^{a_{su}-1} \exp\left(-\frac{x}{b_{su}} - \frac{1}{2}\frac{-2\tau x(\xi_t - (1-\tau_t)x^*) + (\tau_t x)^2}{\tau_t(1-\tau_t)/(\sigma^2(u-t))}\right) \mathrm{d}x}$$

$$= \delta e^{-r_t^u(u-t)} \frac{\int_{\mathbb{X}} x^{a_{su}} \exp\left(-\frac{1}{2}\frac{(\tau x)^2 - 2\tau_t x\,(\xi_t - (1-\tau_t)x^*) + 2\frac{x}{b_{su}}\frac{\tau_t(1-\tau_t)}{\sigma^2(u-t)}}{\tau_t(1-\tau_t)/(\sigma^2(u-t))}\right) \mathrm{d}x}{\int_{\mathbb{X}} x^{a_{su}-1} \exp\left(-\frac{1}{2}\frac{(\tau_t x)^2 - 2\tau_t x\,(\xi_t - (1-\tau_t)x^*) + 2\frac{x}{b_{su}}\frac{\tau_t(1-\tau_t)}{\sigma^2(u-t)}}{\tau_t(1-\tau_t)/(\sigma^2(u-t)}\right) \mathrm{d}x}$$

$$= \delta e^{-r_t^u(u-t)} \frac{\int_{\mathbb{X}} x^{a_{su}} \exp\left(-\frac{1}{2}\frac{(\tau_t x)^2 - 2\tau_t x\left(\xi_t - (1-\tau_t)x^* - \frac{1-\tau_t}{b_{su}\sigma^2(u-t)}\right)}{\tau_t(1-\tau_t)/(\sigma^2(u-t))}\right) \mathrm{d}x}{\int_{\mathbb{X}} x^{a_{su}-1} \exp\left(-\frac{1}{2}\frac{(\tau_t x)^2 - 2\tau_t x\left(\xi_t - (1-\tau_t)x^* - \frac{1-\tau_t}{b_{su}\sigma^2(u-t)}\right)}{\tau_t(1-\tau_t)/(\sigma^2(u-t))}\right) \mathrm{d}x}$$

$$= \delta e^{-r_t^u(u-t)} \frac{\int_{\mathbb{X}} x^{a_{su}} \exp\left(-\frac{1}{2}\left(\frac{\tau_t x - \psi_t}{\gamma_t}\right)^2\right) \mathrm{d}x}{\int_{\mathbb{X}} x^{a_{su}-1} \exp\left(-\frac{1}{2}\left(\frac{\tau_t x - \psi_t}{\gamma_t}\right)^2\right) \mathrm{d}x}, \tag{4.35}$$

where $\psi_t := \xi_t - (1-\tau_t)x^* - \frac{1-\tau_t}{b_{su}\sigma^2(u-t)}$ and $\gamma_t := \frac{\sqrt{\tau_t(1-\tau_t)}}{\sigma\sqrt{(u-t)}}$. A double change of variable, i.e.,

$$\tilde{S}_t = \delta e^{-r_t^u(u-t)} \frac{\int_{\mathbb{X}'} \left(\frac{\gamma_t x' + \psi_t}{\tau_t}\right)^{a_{su}} \exp\left(-\frac{1}{2}(x')^2\right) \frac{\gamma_t}{\tau_t} \mathrm{d}x'}{\int_{\mathbb{X}'} \left(\frac{\gamma_t x' + \psi_t}{\tau_t}\right)^{a_{su}-1} \exp\left(-\frac{1}{2}(x')^2\right) \frac{\gamma_t}{\tau_t} \mathrm{d}x'}, \tag{4.36}$$

with $x' := (\tau_t x - \psi_t)/\gamma_t$ and $\mathbb{X}' = (-\psi_t/\gamma_t, \infty)$, followed by

$$\tilde{S}_t = \delta e^{-r_t^u(u-t)} \frac{\int_{\mathbb{X}} \left(\frac{x}{\tau_t}\right)^{a_{su}} \exp\left(-\frac{1}{2}\left(\frac{x - \psi_t}{\gamma_t}\right)^2\right) \frac{\gamma_t}{\tau_t}\frac{1}{\gamma_t} \mathrm{d}x}{\int_{\mathbb{X}} \left(\frac{x}{\tau_t}\right)^{a_{su}-1} \exp\left(-\frac{1}{2}\left(\frac{x - \psi_t}{\gamma_t}\right)^2\right) \frac{\gamma_t}{\tau_t}\frac{1}{\gamma_t} \mathrm{d}x}$$

$$= \delta e^{-r_t^u(u-t)} \frac{1}{\tau_t} \frac{\int_{\mathbb{X}} x^{a_{su}} \exp\left(-\frac{1}{2}\left(\frac{x - \psi_t}{\gamma_t}\right)^2\right) \mathrm{d}x}{\int_{\mathbb{X}} x^{a_{su}-1} \exp\left(-\frac{1}{2}\left(\frac{x - \psi_t}{\gamma_t}\right)^2\right) \mathrm{d}x}, \tag{4.37}$$

with $x := \gamma_t x' + \psi_t$ and $\mathbb{X} = (0, \infty)$, reveals that the signal-based price S_t can crisply be expressed as the ratio of two consecutive raw (uncentered) absolute (left-truncated at 0) moments of the normal random variable

$$X \sim \mathcal{N}\left(\psi_t, \gamma_t^2\right), \tag{4.38}$$

where, again,

$$\psi_t := \xi_t - (1-\tau_t)x^* - \frac{1-\tau_t}{b_{su}\sigma^2(u-t)} \quad \text{and} \quad \gamma_t := \frac{\sqrt{\tau_t(1-\tau_t)}}{\sigma\sqrt{(u-t)}}. \tag{4.39}$$

With reference to, e.g., [22], Eq. (4.37) can be rephrased more even neatly as

$$\begin{aligned}\tilde{S}_t &= \delta e^{-r_t^u(u-t)} \frac{1}{\tau_t} \frac{\gamma_t^{a_{su}} 2^{\frac{a_{su}}{2}} \frac{\Gamma\left(\frac{a_{su}+1}{2}\right)}{\sqrt{\pi}}}{\gamma_t^{a_{su}-1} 2^{\frac{a_{su}-1}{2}} \frac{\Gamma\left(\frac{a_{su}}{2}\right)}{\sqrt{\pi}}} \frac{{}_1F_1\left(-\frac{a_{su}}{2}, \frac{1}{2}; -\frac{1}{2}\left(\frac{\psi_t}{\gamma_t}\right)^2\right)}{{}_1F_1\left(-\frac{a_{su}-1}{2}, \frac{1}{2}; -\frac{1}{2}\left(\frac{\psi_t}{\gamma_t}\right)^2\right)} \\ &= \delta e^{-r_t^u(u-t)} \tau_t^{-1} \gamma_t \sqrt{2} \frac{\Gamma\left(\frac{1+a_{su}}{2}\right)}{\Gamma\left(\frac{a_{su}}{2}\right)} \frac{{}_1F_1\left(-\frac{a_{su}}{2}, \frac{1}{2}; -\frac{1}{2}\left(\frac{\psi_t}{\gamma_t}\right)^2\right)}{{}_1F_1\left(-\frac{a_{su}-1}{2}, \frac{1}{2}; -\frac{1}{2}\left(\frac{\psi_t}{\gamma_t}\right)^2\right)},\end{aligned} \tag{4.40}$$

where ${}_1F_1(\kappa, \nu; z)$ corresponds to the confluent hypergeometric function of the first kind (or Kummer's function) that is given by

$${}_1F_1(\kappa, \nu; z) \equiv \sum_{m=0}^{\infty} \frac{(\kappa)_m}{(\nu)_m} \frac{z^m}{m!} \tag{4.41}$$

with $(\kappa)_m$ being the Pochhammer symbol defined by

$$(\kappa)_m \equiv \begin{cases} 1, & \text{if} \quad m = 0, \\ \kappa(\kappa+1)\ldots(\kappa+m-1), & \text{if} \quad m > 0. \end{cases} \tag{4.42}$$

Equation (4.41) is known to converge for any $z \in \mathbb{C}$ and is defined for any $\kappa \in \mathbb{C}$, $\nu \in \mathbb{C} \setminus \{\mathbb{Z}^- \cup \{0\}\}$, with $\mathbb{Z}^-$ being the set of negative integers. We also note that ${}_1F_1(\kappa, \nu; 0) = 1$ for all feasible κ, ν. Further details on this type of functions are provided in [21]. Furthermore, in [4], the authors reach a closed-form result in terms of a finite sum of Legendre-type polynomials that is somewhat analogous to Eq. (4.40).

There is in fact a range of fast and effective algorithms available in the literature (see, e.g., [21]) to compute ${}_1F_1(\kappa, \nu; z)$, such as Taylor series, single fraction, Buchholz polynomials, asymptotic series expansion, quadrature methods, or via solving the confluent hypergeometric differential equation (CHDE):

$$z\frac{\mathrm{d}^2 f}{\mathrm{d}z^2} + (\nu - z)\frac{\mathrm{d}f}{\mathrm{d}z} - \kappa f = 0. \tag{4.43}$$

A thorough survey of algorithms that deal with confluent hypergeometric functions is beyond the scope of this chapter, but Taylor series expansion seems to stand out as the most simple and least costly method to compute Eq. (4.41). Picking an appropriate tolerance level, say $e = 10^{-15}$, and introducing, based on Eq. (4.41), the series

$$A_m := \frac{(\kappa)_m}{(\nu)_m}\frac{z^m}{m!}, \qquad \hat{F}_m := \sum_{m=0}^{\infty} A_m, \tag{4.44}$$

with $A_0 = 1$, $\hat{F}_0 = A_0$, and

$$A_{m+1} = A_m\left(\frac{\kappa+m}{\nu+m}\right)\left(\frac{z}{m+1}\right), \qquad \hat{F}_{m+1} = \hat{F}_m + A_m, \qquad \hat{F}_\infty = {}_1F_1, \tag{4.45}$$

the desired function ${}_1F_1$ can easily be computed to a high precision using the following truncation procedure:

$$\hat{F}_M = \sum_{m=0}^{M} A_m, \qquad \text{such that} \quad \frac{|A_{M+1}|}{|\hat{F}_M|} < e. \tag{4.46}$$

This method indeed yields the desired values of ${}_1F_1$ in a small fraction of a second. Figure 4.5 shows the ratio of two confluent hypergeometric functions for several values of κ and z, calculated based on the above method.

Thus, all in all, we are able to recover a crisp tractable approximation formula for the signal-based price of a risky asset at time t which will pay an implied dividend of $\phi(X_u)$ at time u. For computational purposes, we finally note from Eq. (4.40) that,

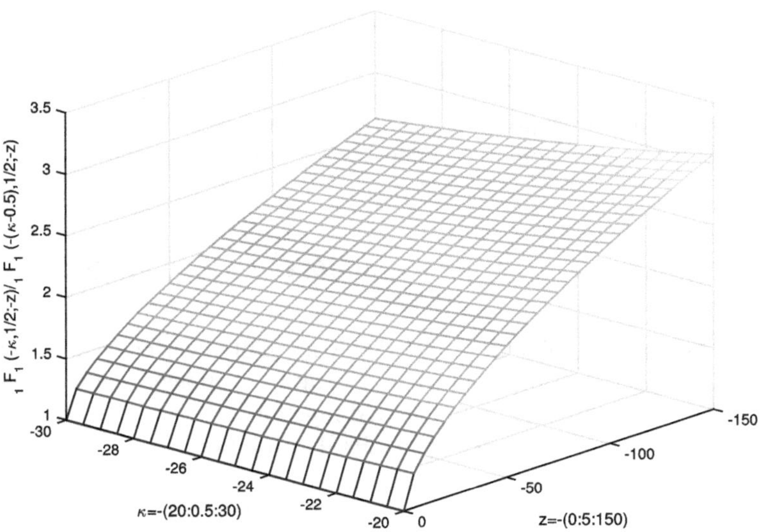

Fig. 4.5 Ratio of two confluent hypergeometric functions whereas values are calculated using Taylor expansion with a tolerance level $e = 10^{-15}$

when $s < t \leq u$ (i.e., with a_s^u and b_s^u having already been inferred from the data) only the last argument of ${}_1F_1$ needs to be updated with the arrival of new information ξ_t – which is expected to improve the algorithm's speed.

4.5.1 Extension to Multiple Signals

At any time t, there will be a total of $k = 1, \ldots, n(t)$, earnings signals, with each of them being τ_k into their lifetime. Thus, the approximate price $\tilde{S}_t$ in the multiple cashflow case is the sum of information-based net present values $\tilde{S}_t^1, \ldots, \tilde{S}_t^{n(t)}$, of a strip of $n(t)$ cashflows, and a Gordon continuation value in the sense of Sect. 4.2.1 above, i.e.,

$$\begin{aligned}\tilde{S}_t &= \delta\left(\sum_{k=1}^{n} e^{-r_t^k(T_k-t)}\left(\phi_t(X_{T_k}) + \mathbf{1}_{\{k=n\}}\frac{\phi_t(X_{T_{k+1}})}{r_b-\mu_0}\right)\right)\\ &= \delta\left(\sum_{k=1}^{n} e^{-r_t^k(T_k-t)}\left(\phi_t(X_{T_k}) + \mathbf{1}_{\{k=n\}}\frac{\phi_t(\phi_{T_k}(X_{T_{k+1}}))}{r-\mu_0}\right)\right)\\ &= \delta\left(\sum_{k=1}^{n} \tilde{S}_t^k\left(1 + \mathbf{1}_{\{k=n\}}\frac{e^{\mu_0 \mathrm{d}T_k}}{r_b-\mu_0}\right)\right),\end{aligned} \tag{4.47}$$

where $r_t^k \neq r_b$, $r_b > \mu_0$, each $\tilde{S}_t^k$ as given in Eq. (4.40) above, and with

$$\phi_t(X_v) = \phi_t(\phi_u(X_v)) \quad (t \leq u \leq v) \tag{4.48}$$

following from the tower property given the definition $\phi_t(X_u) := \mathbb{E}_t[\phi(X_u)]$, or $\mathbb{E}[\phi(X_u)|\xi_t]$. In the next section, we calibrate our earnings model to actual data.

4.5.2 Maximum-Likelihood Estimation of Earnings Model

We recall from Sect. 4.5 that $Y = \Delta \log X$ is normally distributed with $\tilde{\mu}$ and $\tilde{\sigma}$, given in Eqs. (4.24) and (4.25), which are, in turn, functions of the parameters α, μ_0, σ_X and σ_μ. We write the log-likelihood function $\mathcal{L}(\alpha, \mu_0, \sigma_X, \sigma_\mu|\mathbf{y})$, based on the transition density of $\log X$, to be maximised as follows:

$$\begin{aligned}\mathcal{L} &= \mathcal{L}(\alpha, \mu_0, \sigma_X, \sigma_\mu|\mathbf{y})\\ &= \sum_{l=3}^{w+1}\log\left(\frac{1}{\sqrt{2\pi}\tilde{\sigma}_{l-1,l}\sqrt{\Delta T_l}}\exp\left[-\frac{1}{2}\frac{(y_{l-1,l}-\tilde{\mu}_{l-1,l}\Delta T_l)^2}{\tilde{\sigma}_{l-1,l}^2\Delta T_l}\right]\right)\end{aligned}$$

$$
\begin{aligned}
&= \sum_{l=3}^{w+1} \log\left(2\pi\left[\sigma_X^2 \Delta T_l + \frac{\sigma_\mu^2}{2\alpha}\Delta T_l^2(1-e^{-2\alpha\Delta T_{l-1}})\right]\right)^{-1/2} \\
&\quad -\frac{1}{2}\sum_{l=3}^{w+1}\left(\frac{\left(y_{l-1,l} - \left[\mu_{l-2,l-1}^X e^{-\alpha\Delta T_{l1}} + \mu_0(1-e^{-\alpha\Delta T_{l-1}}) - \frac{\sigma_X^2}{2}\right]\Delta T_l\right)^2}{\sigma_X^2\Delta T_l + \frac{\sigma_\mu^2}{2\alpha}\Delta T_l^2(1-e^{-2\alpha\Delta T_{l-1}})}\right) \\
&= -\frac{w-1}{2}\log(2\pi) - \frac{1}{2}\log\left(\prod_{l=3}^{w+1}\sigma_X^2\Delta T_l + \frac{\sigma_\mu^2}{2\alpha}\Delta T_l^2(1-e^{-2\alpha\Delta T_{l-1}})\right) \\
&\quad -\frac{1}{2}\sum_{l=3}^{w+1}\left(\frac{\left(y_{l-1,l} - \left[\mu_{l-2,l-1}^X e^{-\alpha\Delta T_{l-1}} + \mu_0(1-e^{-\alpha\Delta T_{l1}}) - \frac{\sigma_X^2}{2}\right]\Delta T_l\right)^2}{\sigma_X^2\Delta T_l + \frac{\sigma_\mu^2}{2\alpha}\Delta T_l^2(1-e^{-2\alpha\Delta T_{l-1}})}\right),
\end{aligned}
\tag{4.49}
$$

where $\Delta T_l := T_l - T_{l-1}$ and w is the estimation window size (i.e., number of Y samples at each iteration). Indeed, one can easily verify that the function $\mathcal{L}$ is concave.

Log-likelihood calibration procedures for a two-layer stochastic asset pricing model with latent growth parameter (or volatility factor) are not very explicit in the literature, at least to the author's knowledge, and possesses some challenges. In [1], for instance, authors develop a maximum-likelihood calibration method for a two-layer stochastic volatility model where option prices are inverted to produce an estimate of the unobservable volatility state variable. Our GBM model with OU drift, as given in Eqs. (4.5) and (4.6), can also be considered within this difficulty category. The issue with estimating the parameters of our earnings model is that a mean-reverting drift is not directly observable, which can lead to a distortion of parameter estimations, particularly of α.

We therefore replace the unobservable $\mu_{l-2,l-1}^X$, $l \le n$, which goes into Eq. (4.49) with its empirical proxy $\hat{\mu}_{l-2,l-1}^X$ as follows. By Eq. (4.22),

$$
\mu_{l-2,l-1}^X = \frac{\log\left(\frac{X_{l-1}}{X_{l-2}}\right) - \sigma_X W_{\Delta T_{l-2}}}{\Delta T_{l-2}} + \frac{\sigma_X^2}{2}. \tag{4.50}
$$

Thus, when $\operatorname{sgn}(\log X_{l-1}/X_{l-2}) = +1$, $\log X_{l-1}/X_{l-2}$ we replace $\mu_{l-2,l-1}^X$ by its empirical proxy

$$
\hat{\mu}_{l-2,l-1}^{X+} = \frac{\mathbb{E}\log\left(\frac{X_{l-1}}{X_{l-2}}\right)^+}{\Delta T_{l-1}} + \frac{\sigma_X^2}{2} \tag{4.51}
$$

and, when $\text{sgn}(\log X_{l-1}/X_{l-2}) = -1$, by

$$\hat{\mu}^{X-}_{l-2,l-1} = \frac{\mathbb{E}\log\left(\frac{X_{l-1}}{X_{l-2}}\right)^{-}}{\Delta T_{l-1}} + \frac{\sigma_X^2}{2}. \tag{4.52}$$

We considered expected values in Eqs. (4.51) and (4.52) so as to prevent noise from disturbing the estimation of α. The equivalent procedures

$$\arg\max_{\alpha,\beta,\sigma_X,\sigma_\mu} \mathcal{L}(\alpha, \beta, \sigma_X, \sigma_\mu|\mathbf{y}) \tag{4.53}$$

and its necessary first-order optimality conditions

$$\frac{\partial \mathcal{L}}{\partial \alpha} = \frac{\partial \mathcal{L}}{\partial \beta} = \frac{\partial \mathcal{L}}{\partial \sigma_X} = \frac{\partial \mathcal{L}}{\partial \sigma_\mu} = 0 \tag{4.54}$$

then yield the desired results. To illustrate, we estimate the earnings model on selected tickers for the period 2000Q1–2015Q1 using more than 60 quarters of earnings data for each. The model output for each ticker is depicted against the actual earnings data in Fig. 4.6, where the calibrated parameters are reported as figure titles. It can be inferred from the figure that earnings growth is generally

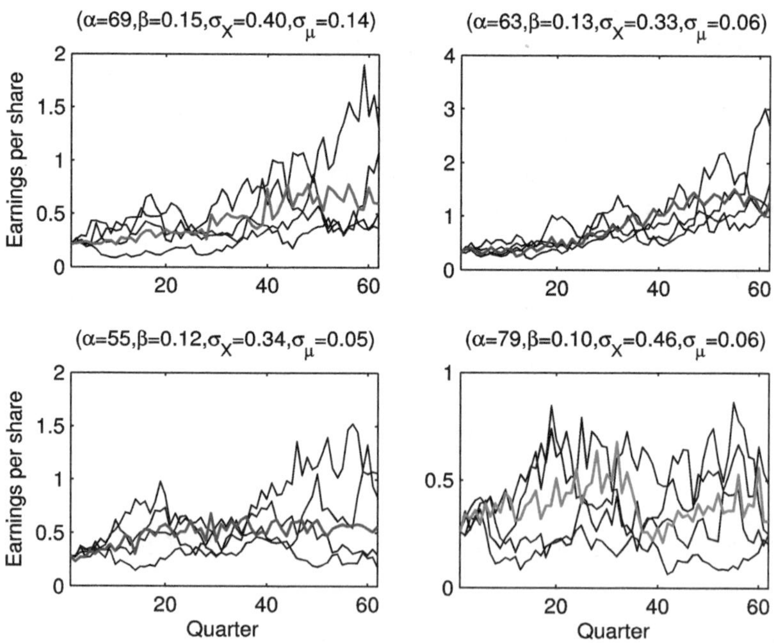

Fig. 4.6 Sample paths of actual earnings (*solid lines*) compared to the calibrated earnings model output (with parameters in headers)

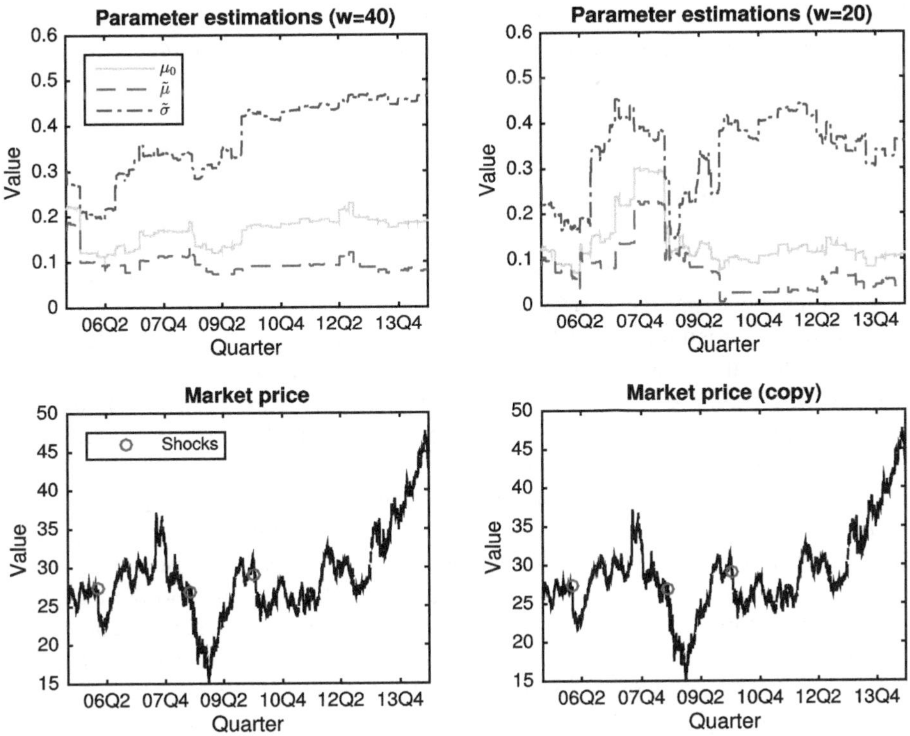

Fig. 4.7 Maximum likelihood parameter estimation of stochastic drift model for implied dividends (*top panel*) and market price (*bottom panel*, two copies to ease vertical comparison)

characterised by large diversions from, as well as extremely fast reversions to, a long-term growth trajectory.[9]

For pricing purposes (in forthcoming Sect. 4.5.3), we shall recursively estimate the parameters of $\mathcal{L}$ using various rolling window lengths w by incorporating both past information and filtered future signals. To illustrate, if the number of available signals at a certain time step t is n_t, the estimation window will then comprise $w - n_t$ and n_t past and future earnings data, respectively. Figure 4.7 depicts the values over time of log-likelihood calibrated parameters, namely, $\tilde{\mu}$, $\tilde{\sigma}$ and μ_0, for the ticker `MSFT` considered in this study (top panels), along with (two copies of) the observed market price for the same period (bottom panels) where major financial incidents are also indicated. Estimated values for α, on the other hand, lie

[9]Alternatively, similar to, e.g., [9], where authors discuss the calibration of stochastic volatility models, $\mu^X_{0,1}$ can be added as an additional parameter to the maximisation problem in Eq. (4.53). Yet, this did not have any significant impact on our results.

in the band [54.8, 191.0]. One notable observation from Fig. 4.7 could be that the estimated model parameters are able to capture major idiosyncratic and systemic incidents of financial stress.

4.5.3 *Information-Based Model Output*

The confluent hypergeometric functions which allowed us to derive a closed-form formula for the signal-based price in terms of Pochhammer series appear rarely in the financial mathematics literature and are generally used as a tool to derive the characteristic function of an average F-distribution as part of the general theory of asset pricing (see, e.g., [15]). In [3], a confluent hypergeometric function appears in the computation of the Laplace transform of the normalised price for arithmetic Asian options. Computation of the confluent hypergeometric functions can pose, however, significant challenges, particularly, when $|z| \gg 0$ (see, e.g., [3, 21]).

For each time step t, we require at least a minimum number of signals be present for the forward-looking information to have sufficient impact on price movements. Figure 4.8, in this respect, shows the number of active signals and their average length for the time period covered in our dataset. Notably, some signals commence as early as over 4 years before their associated earnings are announced. Finally, for $r(t, k)$, i.e., the discount rate, we adopt U.S. T-bill yield curve rates with maturities corresponding (or falling close enough) to that of the cashflow k, $k = 1, \ldots, n(t)$.

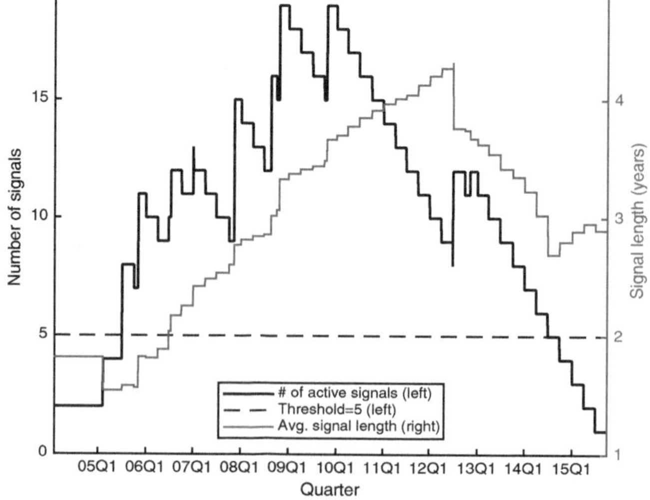

Fig. 4.8 Number n_t (*left*) and average length $T_k - t_k$ (*right*) of active signals over time

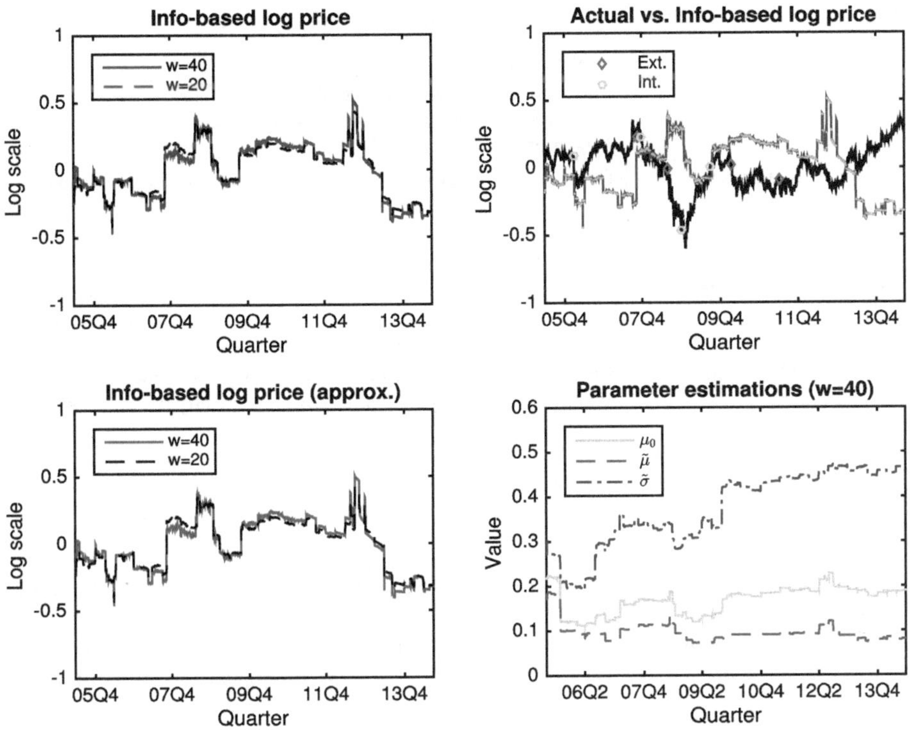

Fig. 4.9 Signal-based price based on multiple signals on quarterly earnings

We accommodate ξ_t for pricing in Eqs. (4.28), (4.34), (4.40) along with Eq. (4.47) to compute both signal-based price S_t (i.e., using (4.28)) and its numerical as well as closed-form approximations $\tilde{S}_t$ (i.e., using Eqs. (4.34) and (4.40), respectively). Figure 4.9 left panels depict the log of the calculated price process (which is also linearly detrended) during the pricing sample period July 22, 2005–October 21, 2014, covering a total of 3379 data points.

Accordingly, we make some immediate observations as follows:

- The numerical results are almost identical to those obtained by the analytical approximation (left panels of Fig. 4.9).
- Since the bulk of the price accumulates the continuation value, which in turn depends on the filtered value of the last cashflow $X_{n(t)}$, the signal-based price is most sensitive to the fluctuations in the last earnings "within" the horizon. This is represented by large swings in the signal-based price, when $t = t_n + 1$ or $t = T_n + 1$.
- Also when $t = T_k$, the contribution of X_k to S simply changes by the amount of surprise (i.e., how much the signal k is off-target just prior to the release of a true factor value). But, more importantly, the surprise at each T_k is incorporated

Table 4.1 Notable reactions of signal-based price to select idiosyncratic and systemic shocks

Date (shock)	Notes
Apr. 27, 2006 (internal)	Although there is no known systemic shock, the signal-based fundamental value quickly reflects the diminishing business growth prospects implied by an unexpected earnings decline.
Dec. 2007 (external)	This is when an across-the-board slowdown in financial activity has started. Yet, there is no significant reaction by the signal-based price, in line with the fact that the real business is yet to be affected.
Sep. 15, 2008 (external)	Lehman collapse. Again, the signal-based price foregoes any significant reaction, until the second round effects hit company's long-term earnings growth prospects.
Jan. 22, 2009 (internal)	Systemic risks starts to threaten business growth outlook (i.e., second round effects), signalled by significantly off-the target earnings.
May 6, 2010 (external)	Known as the "Flash Crash." Again the signal-based price keeps its focus at long-term prospects.

into the signal-based price through improved or deteriorated long-term growth prospects μ_t^0.

- The reaction of the signal-based price to shocks of different types (marked in the top-right panel of Fig. 4.9) has some noteworthy characteristics which are summarised in Table 4.1.

Thus, in this chapter, we availed the signal-based framework for practical use by adapting it to a certain choice of real-time signals.

References

1. Aït-Sahalia Y, Kimmel R (2007) Maximum likelihood estimation of stochastic volatility models. J Financ Econ 83:413–452
2. Bakshi G, Chen Z (2005) Stock valuation in dynamic economies. J Financ Mark 8:111–151
3. Boyle P, Potapchik A (2006) Application of high-precision computing for pricing arithmetic asian options. In: Trager B, Saunders D, Dumas J-G (eds) Proceedings of the international symposium on symbolic and algebraic computation (ISSAC) 2006, Genoa, Italy, July 9–12, ISBN 1-59593-276-3, ACM, pp 39–46
4. Brody D, Hughston LP, Macrina A (2007) Beyond hazard rates: a new framework for credit-risk modelling, In: Advances in mathematical finance, applied and numerical harmonic analysis, chapter III. Birkhäuser, Boston, pp 231–257
5. Brody D, Hughston L, Yang X (2013) On the pricing of storable commodities, Cornell University Library ArXiv e-prints: 1307.5540
6. Campbell J, Shiller R (1988) Stock prices, earnings, and expected dividends. J Financ 43(3):661–676
7. Dong M, Hirshleifer D (2005) A generalized earnings valuation model. Manch Sch 73(Supplement s1):1–31

8. Eisdorfer A, Giaccotto C (2014) Pricing assets with stochastic cash-flow growth. Quant Finan 14(6):1005–1017
9. Fatone L, Mariani F, Recchioni M, Zirilli F (2014) The calibration of some stochastic volatility models used in mathematical finance. Open J Appl Sci 4:23–33
10. Gordon M (1962) The investment, financing, and valuation of the corporation. R.D. Irwin, Homewood
11. Gordon M, Shapiro E (1956) Capital equipment analysis: the required rate of profit. Manag Sci 3(1):102–110
12. Hastie T, Tibshirani R, Friedman J (2009) The elements of statistical learning: data mining, inference, and prediction. Springer series in statistics, 2nd edn. Springer, New York
13. Hoyle A (2010) Information-based models for finance and insurance. Ph.D. Thesis, Department of Mathematics, Imperial College London, London
14. Hoyle A, Hughston L, Macrina A (2011) Lévy random bridges and the modelling of financial information. Stoch Process Appl 121:856–884
15. Hwang S, Satchell S (2012) Some exact results for an asset pricing test based on the average F distribution. Theor Econ Lett 2:435–437
16. Kronimus A (2003) Firm valuation in a continuous-time SDF framework, March 2003, available at: http://www.cofar.uni-mainz.de/dgf2003/paper/paper4.pdf
17. Kullback S, Leibler R (1951) On information and sufficiency. Ann Math Stat 2(1):79–86
18. Lintner J (1956) Distribution of incomes of corporations among dividends, retained earnings, and taxes. Am Econ Rev 76:97–118
19. Longstaff F (2009) Portfolio claustrophobia: asset pricing in markets with illiquid assets. Am Econ Rev 99:1119–1144
20. Longstaff F, Piazzesi M (2004) Corporate earnings and the equity premium. J Financ Econ 74:401–421
21. Pearson J (2009) Computation of hypergeometric functions. Ph.D. Thesis, University of Oxford
22. Winkelbauer A (2012) Moments and absolute moments of the normal distribution, arXiv preprint:1209.4340
23. Yang X (2013) Information-based commodity pricing and theory of signal processing with Lévy information. Ph.D. Thesis, Department of Mathematics, Imperial College London and Shell International, London

Chapter 5
Conclusion

In Chap. 2, we have recovered some of the useful properties of the information-based framework introduced in [2]. This included, inter alia, that the signal process $(\xi_t)_{0\le t\le T}$ was indeed Markov w.r.t. its own filtration and, more strongly, it was dynamically consistent. The latter meant that two agents which observed ξ_t starting from two different time points, say $0,s$, for $s > 0$, would not only have a common view of how ξ_t could evolve in the future (Markov property). They would also have a common view of how X_T could turn out, although the filtration of agent who started observing ξ_t at s was regarded as being generated by $(\xi'_t)_{s\le t\le T}$ instead of $(\xi_t)_{0\le t\le T}$, provided that his a priori knowledge about the terminal law of X_T was updated to $\pi_t(s)$. Furthermore, although the martingale driver W_t was not imposed on the model at the outset, it popped up rather naturally in the price process as a 'reducible' component. It was also shown that, although a higher σ would ensure a less certainty 'at the end' of a certain period about the true fundamental value, a higher σ also meant an elevated price volatility 'during' that period (which seemed somewhat paradoxical) as information was incorporated rapidly. The availability of an exponential martingale for a shift from $\mathbb{Q}$ to $\mathbb{B}$, on the other hand, brought a significant deal of simplification to the problem of derivative pricing. The calculated option prices were indeed in line with the decreasing conditional entropy of (or, uncertainty about) the market factor X_T w.r.t. ξ_t both in time and for growing values of signal-to-noise σ.

In Chap. 3, where a network of a pair of agents with heterogeneous informational skills was introduced, we have seen that the dispersion of the P&L results among agents was directly linked to whether information was revealed through price quotes. The case where agents were 'attentive' and did learn from each other, as compared to the case where they were 'omitters,' was associated with a shrinking of opportunities for (chances of) profit (loss). It was also apparent from the analysis on the impact of learning on the evolution of individual information that the learning process, through updating of posteriors π^j_t, worked in favour of the agent with an inferior individual signal when $\sigma_1 \neq \sigma_2$, and the agents benefited equally otherwise.

N.S. Aydın, *Financial Modelling with Forward-looking Information*, Contributions to Management Science, DOI 10.1007/978-3-319-57147-8_5

As a result, the existence of a common knowledge of gains from trade in the sense of [1] was essential to an equilibrium in the presence of informational asymmetries, and to avoid market shutdowns. For the case where each agent deemed his own signal superior, we have derived explicit formulae for the expected trade signal quality and the potential profits/losses that the agent could make/incur (given his signal pointed at the right/wrong direction), and, thereby, his overall expected P&L before an auction took place. As expected, perception of a greater informational superiority, $|\sigma_1 - \sigma_2|$, meant a greater likelihood for the agent that his trading signal was directionally correct, i.e., $\xi_t^j = \xi_c$, and greater expected profits (vice versa). And this likelihood was stronger in the case of an a priori greater dispersion of the uncertain outcome X_T, and also when the agent chose to refrain from trade. In equilibrium, we found that the optimal strategy was to exploit extra information as it arrived, as the cost of foregoing a profit was higher than the cost of sharing the extra information.

In Chap. 4, we have shown, through a particular example, that the information process and information-based framework can be practically viable, and an analytical approximation to the numerical asset price be recovered. Introducing a slightly modified version of ξ_t and using quarterly earnings consensus data as a basis for constructing the required signals empirically, we approximated the numerical price process via confluent hypergeometric functions of the first kind (or, Kummer's function) in terms of a summation of Pochhammer functions. The model output was notable in that the signal-based price was in general able to capture major trends in the actual price, but it was also successively more responsive to the shocks that were related to the long-term fundamental value of the underlying business, than those that had limited or no impact on the latter.

As an outlook, the present research can be extended in several directions. How a time-varying flow rate σ_t (i.e., agents deem their signal superior only temporally) would affect the equilibrium strategy and P&Ls of agents in Chap. 3 would be an interesting issue to look into. Moreover, making the amount of information shared a function of the amount traded would give the agents the additional flexibility of deciding 'how much information to share,' in addition to 'when to share,' and possibly affect their trading strategies $(q_t^j)_{0\leq t\leq T}$. Finally, the analysis in Chap. 4 reveals that abrupt price changes do actually result from sudden changes in the amount and shape of available information. This allows to extend the analysis in this chapter to a more realistic case by using Lévy processes to model ξ_t.

5.1 Financial Signal Processing (FSP)

The use of digital signal processing (DSP) techniques in financial modelling as a method at the core of engineering discipline is becoming increasingly widespread. FSP, as an branch of DSP, applies techniques from the latter to aid quantitative investment strategies. The overall aim of the theory of FSP is to construct optimal casual filters to extract useful information from a broad range of financial signals.

In the financial context, a signal can be deemed to be the price, or any other, process sampled at a certain frequency which has a certain degree of explanatory power on the variable of interest.

The justification for the use of signal processing techniques for modelling financial data stems from the simple fact that any process in the time domain can be expressed as an ensemble of infinite sinusoidal cycles, each characterised by a distinct cyclic or angular frequency and radius in the frequency domain. Finite impulse-response (FIR) filters, in this regard, are generally preferred due to their stability, linear phase response, flexibility in shaping the magnitude response, and convenience in implementation.

One of the most potent questions pertaining to the application of DSP techniques to finance is about how to deal with latency without trading off attenuation of noise in a causal filter context. This involves designing of, e.g., FIR, filters with selectively prescribed delays in specific frequency regions without adversely influencing the attenuation. This requires a methodology that would take the desired specifications in amplitude, phase or group delay over a band of frequencies, and deliver the required transfer function. One feasible approach is to use root moments, as described in [6]. Hilbert transform is also a useful tool to move from amplitude to phase, so as to achieve the objective of minimising the phase delay.

There are basically two separate issues involved forecasting that need to be dealt with separately, namely, signal 'representation' and 'signal prediction.' Existing techniques, in the main, focus the second issue and consider the first as given and compliant. The 'surrogate signal method,' on the other hand, as proposed in [4], emerges from the basic idea that the latter two problems must be decoupled from each other, and an efficient representation of the signal must precede, and be the basis for, its prediction. In this respect, the surrogate signal, which aims to offer a satisfactory representation of the original signal, is derived from the latter in a way that it retains the desirable attributes of the parent signal, while also satisfying a priori external and equally desirable constraints, such as smoothness and predictability. One particular way to extract the surrogate is through the use of 'annihilator.' Extracted surrogates are linked to trading decisions through a quality factor, and specification of a surrogate quality threshold.

The identification of dominant cycles, i.e., the peak in the representation of the signal in the frequency-amplitude plane through Fourier transform (signal spectrum), is another important concept in DSP. This component is sometimes used to develop momentum as well as high-frequency trading strategies. For non-stationary signals, however, the dominant cycle is generally time-varying and needs to be detected recursively. This gives rise to the issue of instantaneous frequency (as an alternative to filter bank) and the necessity of adaptive filtering techniques (cf. [4]).

Another point where sophisticated DSP techniques can be of great help is basically by introducing the concept of 'smooth independent components,' which implies that the independent components resulting from the independent component analysis (ICA), a well-known blind source separation algorithm, can be constructed in a way that they are robust and stable and, therefore, applicable to maximum portfolio diversification. One example to this is given in [5], where the smooth ICA is used to compactly represent a portfolio of assets.

Finally, the first difference or natural logarithm are generally used as the customary starting to ensure stationarity in financial data, although they sometimes reduce the information component. There are some recent techniques, such as empirical data decomposition (EMD) and the like, which do not require a resort to such transformations while preserving some of the desired characteristics of the data (cf. [3]).

References

1. Bond P, Eraslan H (2010) Information-based trade. J Econ Theory 145:1675–1703
2. Brody D, Hughston LP, Macrina A (2007) Beyond hazard rates: a new framework for credit-risk modelling. In: Advances in mathematical finance, applied and numerical harmonic analysis, chapter III. Birkhäuser, Boston, pp 231–257
3. Chanyagorn P, Cader M, Szu H (2005) Data-driven signal decomposition method. In: Proceedings of the 2005 IEEE international conference on information acquisition
4. Constantinides A (2015) Financial signal processing: a new approach to data driven quantitative investment. In: Presentation to 2015 IEEE international conference on digital signal processing (DSP), 21–24 July 2015, Singapore
5. Korizis H, Mitianoudis N, Constantinides A (2007) Compact representations of market securities using smooth component extraction. In: Davies M, James C, Abdallah S (eds) Independent component analysis and signal separation: 7th international conference, ICA 2007, London, UK, 9–12 September 2007, Springer
6. Stathaki T, Fotinopoulos I, Constantinides A (1999) Root moments: a nonlinear signal transformation for minimum FIR filter design. In: Proceedings of the IEEE-EURASIP workshop on nonlinear signal and image processing (NSIP'99), Antalya, Turkey, 20–23 June 1999

Appendix A
Analytical Gamma Approximation to Log-Normal via Kullback–Leibler Minimisation

We recall the objective function related to Kullback–Leibler distance minimisation problem (4.31):

$$D(a_t, b_t) = \int_{\mathbb{X}} f_X(\tilde{\mu}_t, \tilde{\sigma}_t) \log \left(\frac{f_X(\tilde{\mu}_t, \tilde{\sigma}_t)}{g_X(a_t, b_t)} \right) \mathrm{d}x, \tag{A.1}$$

where $\mathbb{X} = (0, \infty)$. Let $h(\tilde{\mu}_t, \tilde{\sigma}_t)$ denote the terms which don't depend on a_t and b_t. We have

$$D(a_t, b_t) = h(\tilde{\mu}_t, \tilde{\sigma}_t) + \log\left(\Gamma\left(a_t\right)\right) + a_t \log\left(b_t\right) + \frac{1}{b_t}\mathbb{E}_f\left[X\right] - (a_t - 1)\mathbb{E}_f\left[\log\left(X\right)\right]. \tag{A.2}$$

Taking derivatives of D with respect to its arguments, each set to zero, we get

$$\frac{\partial D(a_t, b_t)}{\partial a_t} = \Psi^{(0)}\left(a_t\right) + \log\left(b_t\right) - \mathbb{E}_f\left[\log\left(X\right)\right] = 0 \tag{A.3}$$

$$\begin{aligned} \frac{\partial D(a_t, b_t)}{\partial b_t} &= \frac{a_t}{b_t} - \frac{1}{b_t^2}\mathbb{E}_f\left[X\right] = 0 \\ &= a_t b_t - \mathbb{E}_f\left[X\right] = 0. \end{aligned} \tag{A.4}$$

where

$$\Psi^{(m)}(a_t) \equiv \mathrm{d}^{m+1} \log \Gamma(a_t) / \mathrm{d}a_t^{m+1} \tag{A.5}$$

N.S. Aydın, *Financial Modelling with Forward-looking Information*, Contributions to Management Science, DOI 10.1007/978-3-319-57147-8 is the polygamma function. Knowing that $\mathbb{E}_f[\log(X)] = \tilde{\mu}_t$ and $\mathbb{E}_f[X] = \exp(\tilde{\mu}_t + \tilde{\sigma}_t^2/2)$, we obtain the following system of equations to solve:

$$\Psi^{(0)}(a_t) + \log(b_t) = \tilde{\mu}_t$$
$$a_t b_t = \exp\left(\tilde{\mu}_t + \frac{\tilde{\sigma}_t^2}{2}\right). \tag{A.6}$$

Next we eliminate b_t by inserting first equation into the latter

$$a_t = \exp\left(\Psi^{(0)}(a_t) + \frac{\tilde{\sigma}_t^2}{2}\right). \tag{A.7}$$

A first-degree approximation to $\Psi^{(0)}(a_t)$ is given by

$$\Psi^{(0)}(a_t) \approx \log(a_t) - \frac{1}{2a_t} \tag{A.8}$$

which yields

$$a_t \approx \frac{1}{\tilde{\sigma}_t^2}, \quad b_t \approx \tilde{\sigma}_t^2 \exp\left(\tilde{\mu}_t + \frac{\tilde{\sigma}_t^2}{2}\right). \tag{A.9}$$

FSC
www.fsc.org
MIX
Papier | Fördert gute Waldnutzung
FSC® C083411

Kolibri 360 GmbH
Ferdinand-Jühlke-Straße 7
99095 Erfurt, Deutschland
produktsicherheit@kolibri360.de